Fundamentals of Measurable Dynamics

Fundamentals of Measurable Dynamics

Ergodic Theory on Lebesgue Spaces

DANIEL J. RUDOLPH

Department of Mathematics, University of Maryland

CLARENDON PRESS · OXFORD

1990

Oxford University Press, Walton Street, Oxford OX2 6DP
Oxford New York Toronto
Delhi Bombay Calcutta Madras Karachi
Petaling Jaya Singapore Hong Kong Tokyo
Nairobi Dar es Salaam Cape Town
Melbourne Auckland
and associated companies in
Berlin Ibadan

Oxford is a trade mark of Oxford University Press

Published in the United States
by Oxford University Press, New York

British Library Cataloguing in Publication Data
Rudolph, Daniel J.
Fundamentals of measurable dynamics.
1. Ergodic theory
I. Title
515.42
ISBN 0–19–853572–4

Library of Congress Cataloging in Publication Data
Rudolph, Daniel J.
Fundamentals of measurable dynamics: ergodic theory on Lebesgue spaces / Daniel J. Rudolph.
Includes bibliographical references (p.).
Includes index.
1. Ergodic theory. 2. Measure-preserving transformations.
I. Title.
QA614.R83 1990
515′.43—dc20 90-7486
ISBN 0–19–853572–4

Set by
Asco Trade Typesetting Ltd, Hong Kong.
Printed and bound in Great Britain by
Biddles Ltd, Guildford and King's Lynn

Preface

Our intention here is to give an elementary technical treatment of the fundamental concepts of the measure-preserving dynamics of a Lebesgue probability space. This text has grown out of a course given at the University of Maryland beginning in the Spring of 1984.

The last twenty-five years have seen an enormous growth in the theory of dynamical systems in general, and in particular, the probabilistic side of this field, classically known as ergodic theory. This development in recent years, most especially through the work of D. S. Ornstein and his school, has changed the perspective on much of the classical work in the field to a more set-coding combinatorial point of view as opposed to a functional analytic point of view. This perspective is of course much older, easily visible in the work of Kakutani, Chacón, and many others. We choose to attach Ornstein's name to it as it is in his work, and the work of those around him, that this point of view has reached its current power.

Good expository treatments of ergodic theory already exist and certainly proofs of many of our main results can easily be found in standard texts in the field. What we are attempting to do through the methodology and order of proof we have chosen is to present the reader with this body of material from what has been to date a most fruitful point of view with the fabric of its development, at least at an elementary level, intact.

We assume the reader has a thorough working knowledge of the topology of the real line, and Lebesgue measure theory on the real line. Royden (1968) is a good source for this material.

The one deep result we will use without proof is the Riesz representation theorem for the dual of the continuous functions but only on such simple topological spaces as Cantor sets or the unit circle.

The text is not intended to be encyclopaedic, but rather to present detailed arguments and chains of arguments showing technically how the fundamentals of dynamics on Lebesgue spaces are developed. The intent is to show those who want to prove theorems in ergodic theory what some of the more fruitful threads of argument have been. For this reason, we gladly present some very technical material, and at several points give multiple proofs. Although in places the technical detail may seem formidable, we will in fact often make simplifying assumptions. The one most obvious is that only the ergodic theory of single transformations is considered. To extend the theory to actions of $\mathbb{Z}^n$ or more general discrete abelian groups is not too difficult. For non-abelian and continuous groups, even as basic as $\mathbb{R}$, extension requires new ideas, fundamentally the existence of measurable sections. We include a bibliography where such extensions can be found. From the basis given here, this literature should be readily accessible.

Chapter 1 presents the fundamental concepts of measure-preserving dynamics and introduces a number of examples.

Chapter 2 is a basic and technical treatment of the structure of Lebesgue probability spaces. As a preparation for later work, we prove an L^1-martingale theorem via the Vitali covering lemma. This argument is a warm-up for our proof of both the Birkhoff ergodic theorem and the Shannon–McMillan–Breiman theorem, as both will be proven by a 'Vitali' type argument.

Chapter 3 presents the ergodic theorems, and ergodic decomposition. We present the now classical von Neumann L^2-ergodic theorem and the Garsia–Halmos proof of the Birkhoff ergodic theorem, to juxtapose them with the 'backward Vitali lemma' proof we then present. The intention is to give the reader as much technical insight into these theorems as is reasonable. Our last task is to show that any measure-preserving transformation decomposes as an integral of ergodic transformations. Hopefully the presentation given here makes this very technical argument approachable. Proofs are difficult to find in the literature.

Chapter 4 covers the hierarchy of mixing properties, presenting the circle of definitions of weakly mixing and ending with the definition of a Kolmogorov automorphism. Included is a short development of the spectral theory of transformations.

This leads to the theory of entropy in Chapter 5, which we develop from a name-counting point of view, again using the backward Vitali lemma to prove the Shannon–McMillan–Brieman theorem.

In Chapter 6 we introduce the concept of a joining and disjointness and we use these to again characterize ergodic weakly mixing, and K-mixing transformations. We also show that Chacón's map has minimal self-joinings and use this to construct some counter-examples.

In Chapter 7 we present the Burton–Rothstein proofs of Krieger's generator theorem and Ornstein's isomorphism theorem from the viewpoint of joinings.

Many exercises are presented throughout the course of the text. They are intended to help the reader develop technical facility with the methods developed, and to explicate areas not fully developed in the text.

Chapters 1 through 5 form the core of an introductory graduate course in ergodic theory. As the point of view here is technical, we have been most successful using this material in conjunction with a more broadly oriented text such as Walters (1982), Friedman (1970) or Cornfeld, Formin and Sinai (1982). Chapters 6 and 7 can be used either as the core of a more advanced seminar or reading course, expanded perhaps with appropriate research literature of the field, allowing the instructor to orient the course either toward deeper abstract study, or application of the material. The book ends with a bibliography.

I would like to thank Charles Toll, Ken Berg, Mike Boyle, Aimee Johnson and Janet Kammeyer for collecting and refining the original notes, and the text during its years of development. I also must thank Virginia Vargas for shepherding along the manuscript through its many revisions.

Maryland D. J. R.
1989

Contents

1 Measurable dynamics

Dynamics, as a mathematical discipline, is the study of those properties of some collection of self-maps of a space which become apparent asymptotically through many iterations of the maps. The collection is almost always a semigroup, usually a group, and often, as a group, simply $\mathbb{Z}$ or $\mathbb{R}$. The origins of the field lie in the study of movement through time of some physical system. The space is the space of possible states of the system and the self-maps indicate how the state changes as time progresses.

The field of dynamics breaks into various disciplines according to the category of the space and self-maps considered. If the space is a smooth manifold and the self-maps are differentiable then the discipline is smooth dynamics. If the space is just a topological space and the self-maps are continuous, the field is topological dynamics. As a very significant subfield of this, if the space is a closed shift-invariant subset of all infinite sequences of symbols from some finite set, and the self-maps are the shifts, then the field is symbolic dynamics. Lastly, and of interest to us here, if the space is a measure space and the self-maps are measure-preserving then the field is measurable dynamics, or more classically, ergodic theory.

Each of these disciplines has its own special flavour and language, but the overlaps among them are enormous and the parallels often subtle and insightful.

1.1 Examples

Here is a collection of examples of dynamical systems from various mathematical disciplines, each of which has one or more natural invariant measures.

Example 1 Rotations of the circle. The space S^1 will be the circle of circumference 1 and the self-maps, R_α, rotations by an angle α. In this case, the space and the collection of maps can be identified as the same object, the group of rotations of the circle.

Lebesgue measure m on the circle is invariant under rotations. This is a particular case of a compact group acting on itself by left multiplication and its unique invariant Haar probability measure. We will return to this idea in our last example.

Example 2 Hyperbolic toral automorphisms. Here the space T^2 will be the two-dimensional torus. We think of this as $\mathbb{R}^2/\mathbb{Z}^2$. This is again a compact abelian group but we will not consider the action of the group on itself. Let

$M \in SL(2, \mathbb{Z})$ be a 2×2 integer matrix of determinant 1, (for example $\begin{bmatrix} 2 & 1 \\ 1 & 1 \end{bmatrix}$). As $M(\mathbb{Z}^2) = \mathbb{Z}^2$, M projects to an automorphism of T^2 preserving Lebesgue measure. Suppose the eigenvalues of M are λ, λ^{-1} and do not lie on the unit circle. They both must be real. Let unit eigenvectors be $\mathbf{v}_1$, and $\mathbf{v}_2$. Supposing $|\lambda| < 1$ for any point $x \in T^2$, the line of points $x(s) = x + s\mathbf{v}_1$ contracts exponentially fast as M is iteratively applied, i.e.,

$$\|M^n(x(s)) - M^n(x)\| = |\lambda|^n |s|.$$

Similarly, the line $x(u) = x + u\mathbf{v}_2$ expands exponentially fast. The existence of such expanding and contracting foliations (families of lines or curves), is a common, and much studied, phenomenon. It has powerful implications for the dynamics of the system, as we see in Exercise 2.

Example 3 Mixing Markov Chains. Let M be an $n \times n$ matrix whose entries $m_{i,j}$ are either 0 or 1. Assume that for some $k \geq 0$ all entries of M^k are greater than 0. The space $\sum_M$ will consist of all sequences $\mathbf{x} = \{x_j\}_{j=-\infty}^{\infty}$, $x_j \in \{1, \dots, n\}$ where

$$m_{x_j x_{j+1}} = 1.$$

We view the movement from x_j to x_{j+1} as a transition from one state to another and the 1's in M indicate the allowed transitions. $M^k > 0$ says that all transitions across k time units are allowed. The map $T: \sum_M \to \sum_M$ is the left shift, i.e., $T(\mathbf{x}) = \mathbf{y}$ where $y_i = x_{i+1}$. The space $\sum_M$ is a closed shift-invariant subset of $\{1, \dots, n\}^{\mathbb{Z}}$, and is called a topological Markov shift, or chain. In this situation there are many invariant measures. Here is one sort. Let $P = [p_{i,j}]$ be a matrix with all $p_{i,j} \geq 0$, $p_{i,j} > 0$ iff $m_{i,j} = 1$ and

$$\sum_{j=1}^{n} p_{i,j} = 1,$$

i.e., a Markov matrix compatible with M. As P^k has all positive entries, P has a unique left eigenvector $(p_1, \dots, p_n)$ with eigenvalue 1 with all $p_i > 0$ and $\sum_{i=1}^{n} p_i = 1$.

This allows us to define a T-invariant finitely additive measure on open sets by defining it on a cylinder set $c = \{\mathbf{x} : x_j = i(j), a \leq j \leq b\}$ to be

$$\mu_P(c) = p_{i(a)} \prod_{k=a}^{b-1} p_{i(k), i(k+1)}.$$

That μ_P extends to a T-invariant Borel measure on $\sum_A$ follows from the Kolmogorov extension theorem (see Chapter 2).

Example 4 Isometries of compact metric spaces. Here the space X will be a compact metric space with metric $d(\ ,\ )$, and the self-map $f: X \to X$, an isometry $d(f(x), f(y)) = d(x, y)$.

We will assume f has a dense orbit, i.e., for some x_0

$$\overline{\bigcup_{n=-\infty}^{\infty} f^n(x_0)} = X. \tag{1.1}$$

If f has such a dense orbit, then all positive orbits are dense, i.e., $\overline{\bigcup_{n=1}^{\infty} f^n(x)} = X$ for all $x \in X$. We want to see two things about this situation; first that X can be made a compact abelian group with the action of f given as multiplication by $f(x_0)$ and second that X has a unique f-invariant Borel probability measure.

If X is a finite set, then f must be a cyclic permutation and the results are easy to see.

Thus we may assume that all the points $f^n(x_0)$ are distinct. We first define a product rule on this dense subset of $X \times X$ by $f^n(x_0) \times f^m(x_0) = f^{n+m}(x_0)$. If $f^{n_i}(x_0)$ and $f^{m_i}(x_0)$ are both Cauchy sequences, then so is $f^{n_i - m_i}(x_0)$. Thus for $a, b \in X$, select $f^{n_i}(x_0) \to a$, $f^{m_i}(x_0) \to b$ and define $a \times b^{-1} = \lim f^{n_i - m_i}(x_0)$. This element of X is independent of the Cauchy sequences chosen, and makes X an abelian topological group. The group relations follow from those on $f^n(x_0)$.

Notice that x_0 is the identity element, $f(x_0) \times a = f(a)$ and multiplying by any $a \in X$ is an isometry of X.

Remarks

1. A compact group with an element g with $\{g^n : n \geq 0\}$ dense is called *monothetic*.

2. This argument can be generalized to much weaker situations; for example X need not be metric, only compact T_2 with $\{f^n\}$ equicontinuous. The index n could be replaced by $t \in \mathbb{R}$ or more generally by $g \in G$, a locally compact abelian group of self-maps of X. We could also have started with X merely precompact and extended f to its compactification. We will apply this last idea in our discussion of the weakly mixing property in Chapter 4.

3. If we do not assume f has a dense orbit, then X always decomposes into an indexed family $X(\alpha)$ of disjoint compact f-invariant sets on each of which f has a dense orbit. This is a simple case of the decomposition of a space into 'ergodic components.'

We now want to see that f has a unique invariant Borel probability measure. We will do this by showing that a compact abelian metrizable monothetic group has a unique invariant Haar measure. For any continuous function h, set

$$A_n(h, x) = \frac{1}{n} \sum_{i=0}^{n-1} h(g^i \times x) \tag{1.2}$$

where g has a dense orbit.

Using the density of the orbit of g one shows that

$$\lim_{n\to\infty} |A_n(h,x) - A_n(h,y)| = 0 \tag{1.3}$$

uniformly in $X \times X$, and hence as

$$A_{mn}(h,x) = \frac{1}{m}\sum_{j=0}^{n-1} A_n(h, g^{jm} \times x), \tag{1.4}$$

we can conclude

$$\lim_{n\to\infty} A_n(h,x) = L(h)$$

converges uniformly to a constant for $x \in X$.

As $L(h)$ is a bounded linear functional on $C(X)$, by the Riesz representation theorem

$$L(f) = \int f \, d\mu \tag{1.5}$$

for some Borel measure μ. That μ is g invariant follows easily from

$$L(h) = L(g(h)).$$

If v were some other g-invariant measure, then

$$\int A_n(h,x)v(dx) = \int hv(dx).$$

But as $A_n(h,x)$ converges uniformly to $L(h)$, $v = \mu$, and μ is unique. This is only one of myriad constructions of Haar measure for such groups.

Example 5 Rank-1 cutting and stacking constructions. We will describe a general method for constructing a certain class of maps. Each will act on some interval $[0, a) \subset \mathbb{R}^+$. The method is inductive. At each stage of the induction we will have constructed a partial map f_n, defined on the interval $[0, a_n)$, where $a_n \leq a$. The map f_n will have the following form. The interval $[0, a_n)$ will be cut into disjoint subintervals

$$I(1,n), I(2,n), \ldots, I(N(n),n),$$

left-closed, right-open, all of the same length $a_n/N(n)$. We will have assigned some permutation π_n of $\{1,\ldots,N(n)\}$ to order the intevals and f_n maps $\pi_n(i)$ to $\pi_n(i+1)$ linearly. Thus f_n is defined everywhere except $I(\pi_n(N(n)),n)$, and f_n^{-1} is defined everywhere except $I(\pi_n(1),n)$. We view this situation as a stack, tower or block of intervals (all three words are often used) and f_n as movement vertically through the stack (Fig. 1.1). To get started

$$N(0) = 1, \quad I(1,0) = [0,1), \quad \text{and} \quad \pi_0 = \text{id}. \tag{1.6}$$

To construct f_{n+1}, we choose a parameter $k(n+1) > 1$ and cut each $I(j,n)$

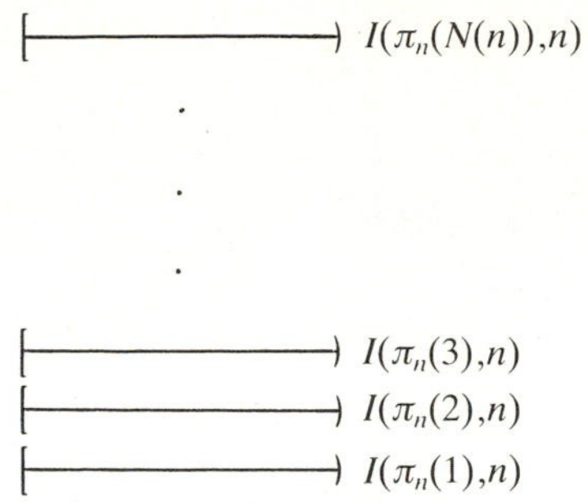

Fig. 1.1 A stack of intervals.

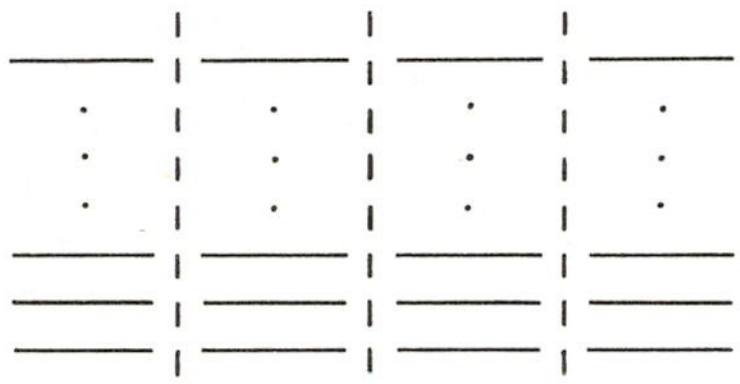

Fig. 1.2 Cutting a stack.

into $k(n+1)$ subintervals of the same width. We can view this as slicing through the stack vertically (Fig. 1.2). This gives us the first part of our list of intervals $I(j, n+1)$. We also select parameters

$$S(1, n+1), S(2, n+1), \ldots, S(k(n+1), n+1) \in \mathbb{Z}^+ \cup \{0\}$$

and cut off

$$\sum_{j=1}^{k(n+1)} S(j, n+1)$$

intervals, of the same length

$$\frac{a_n}{N(n)k(n+1)} \tag{1.7}$$

as those already cut, but from an interval $[a_n, a_{n+1})$.

We can define π_{n+1} by describing how to stack these intervals. We work from left to right through the sliced off subcolumns of the previous stack, placing one above the other, putting $S(j, n+1)$ of our new intervals atop the jth slice before adding the next (Fig. 1.3). $\pi_{n+1}(i)$ is the index of the interval i steps up the stack of intervals. It is easy to see that $f_{n+1} = f_n$ where both are defined, and that all f_n preserve Lebesgue measure ℓ. We assume

$$\sum_{n=1}^{\infty} \frac{\sum_{j=1}^{k(n)} S(j, n) a_{n-1}}{N(n-1)k(n)} = a - 1 < \infty \tag{1.8}$$

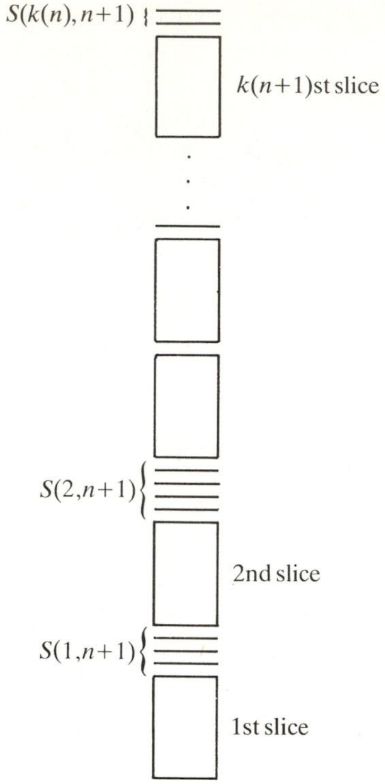

Fig. 1.3 Forming the next stack.

so

$$\lim_{n\to\infty} a_n = a < \infty$$

and if we normalize ℓ to $\mu = \ell/a$, the limit map f preserves the probability measure μ, and is 1–1 onto from $[0, a) \to (0, a)$. Omitting the forward orbit $f^n(0)$, $n \in \mathbb{Z}^+$, we are left with a measure-preserving invertible map. The map f is characterized by the sequence of parameters $k(n)$, $S(1, n), \ldots, S(k(n), n)$ and is called a rank-1 cutting and stacking construction. Some interesting examples are:

(1) $k(n) = 2$, $S(j, n) = 0$ for all j, n, called the dyadic adding machine; and

(2) $k(n) = 3$, $S(1, n) = S(3, n) = 0$, $S(2, n) = 1$, an example due to Chacón that has very interesting properties which we will return to in Chapters 3 and 6.

We have now described a little collection of examples of measure-preserving dynamical systems. We have limited our list to what are the most easily described, or most useful for our later purposes. As we continue we will add other examples and elaborate on those already given.

The fundamental problem of any branch of dynamics is to find methods for classifying dynamical systems up to the category of interest. In the case of ergodic theory the problem is to find methods, or structures, which classify measure-preserving systems which are invariant under measurable isomorphism.

What does it mean for two dynamical systems to be measurably isomorphic? Suppose G is a semigroup and we have two measure spaces $(X, \mathscr{F}, \mu)$ and $(X', \mathscr{F}', \mu')$, and for $g \in G$, measure-preserving maps, composing according to the multiplication rules of G, $T_g : X \to X$, $T'_g : X' \to X'$.

Definition 1.1 We call these systems *isomorphic* if there are invariant subsets $X_0 \subset X, X'_0 \subset X'$, each of full measure, and a measurable measure-preserving invertible map,

$$\phi : X_0 \to X'_0 \tag{1.9}$$

so that

$$\phi T_g = T'_g \phi \quad \text{for all } g \in G.$$

Any properties of interest to us in ergodic theory must be preserved by isomorphism.

For example, the property of *ergodicity*, i.e., that any set A with $T_g^{-1}(A) = A$ for all $g \in G$ must have measure 0 or 1, is an isomorphism invariant. In our list of examples, we can show virtually all the maps discussed were ergodic.

Another property we will verify for some of our examples is *mixing*, that for any measurable sets A and B

$$\lim_{n \to \infty} \mu(T^n(A) \cap B) = \mu(A)\mu(B). \tag{1.10}$$

This is also an isomorphism invariant.

The exercises that follow involve verifying these properties for some of our examples.

1.2 Exercises

Many of these exercises are quite difficult. We recommend the reader keep returning to them while progressing through later chapters. They are not intended so much to test the reader's understanding of the chapter, as to lead the reader more deeply into the material.

1.1 Let R_α be rotation by α on S^1. Suppose R_α has an invariant set, i.e., there is a set $A \subset S^1$, $m(A) \neq 0, 1$ and $R_\alpha(A) = A$. Show that $\alpha = 2\pi r$, $r \in \mathbb{Q}$ (the rationals).

1.2 Let $M \in SL(2, \mathbb{Z})$ with eigenvalues λ, λ^{-1} not on the unit circle. Using the fact that the two-dimensional trigonometric polynomials are dense in $L^2(T^2)$, for $f, g \in L^2(T^2)$ show that

$$\lim_{n\to\infty} \int f\cdot g(M^n)\,\mathrm{d}\mu = \int f\,\mathrm{d}\mu \int g\,\mathrm{d}\mu.$$

Hence if $f \circ M = f$, f is a constant μ–a.e.

1.3 Let $P = [p_{i,j}]$ be a Markov matrix as in Example 3, and $\mathbf{p}$ its unique left eigenvector with eigenvalue 1, as described.

1. Using the ℓ^1 metric $\|\mathbf{v}\| = \sum_{i=1}^n |v_i|$ on $\mathbb{R}^n$, show that if $\sum_{i=1}^n \mathbf{v}_j = 0$ then $\lim \|\mathbf{v}P^n\| = 0$. Hint: Show $\|\mathbf{v}P\| \leq \|\mathbf{v}\|$ and then splitting $\mathbf{v}$ into positive and negative terms show $\|\mathbf{v}P^k\| \leq \lambda\mathbf{v}$ where $\lambda = 1 -$ (smallest entry in P^k)/$2n$.

2. Use part 1 to show that for any probability vector $\mathbf{v}$,

$$\lim_{n\to\infty} \|(\mathbf{p} - \mathbf{v})P^n\| = 0.$$

3. Use part 2 to show that for any cylinder sets C_1, C_2,

$$\lim_{n\to\infty} (\mu_P(C_1 \cap T^n(C_2))) = \mu_P(C_1)\mu_P(C_2).$$

4. Use part 3 to show that for any $f \in L^1(\sum_A, \mu_P)$ if $f \circ T = f$ then $f =$ constant μ_P–a.e.

1.4 Suppose (X, d, f) is as in Example 4. Show that if f has a dense orbit, then all positive orbits of f are dense.

1.5 Work through the details of the argument sketched in the remarks following Example 4.

1.6 Show that if f is a rank-1 cutting and stacking construction and $h \in L^1(\mu)$ then

$$A_n(h, x) = \frac{1}{n}\sum_{i=0}^{n-1} h(f^i(x))$$

converges in $L^1(\mu)$ to $\int h\,d\mu$. (First show this for functions constant on intervals $I(j, n)$ and then use their density in $L^1(\mu)$.)

1.7 Show that no two of the following are isomorphic:
(1) rotation on the circle by an angle $2\pi\alpha$, $\alpha \in \mathbb{Q}$;
(2) a hyperbolic toral automorphism;
(3) the dyadic adding machine (hint: consider T^2h).

2 Lebesgue probability spaces

The examples of measure-preserving dynamical systems of the previous chapter sat on a wide variety of spaces. Some were merely topological, others were smooth manifolds, often groups. All we are fundamentally interested in is the fact that they are measure spaces endowed with a probability measure. The purpose of this chapter is to show that under some easily verified assumptions, all such spaces are measurably conjugate to Lebesgue measure on the unit interval in $\mathbb{R}^1$. Thus the underlying probability space, unless very unusual, will never provide an obstacle to the existence of a measurable isomorphism.

This fact allows us to use, on such 'Lebesgue spaces', all the machinery of real analysis, most especially the Vitali covering lemma. We do so to prove an L^1-martingale theorem at the end of the chapter.

We now describe the basic structure whose existence on a given set and finitely additive measure will imply that, as far as measurable behavior is concerned, we are dealing with the unit interval and Lebesgue measure.

2.1 Countable algebras and trees of partitions

To begin, let X be some arbitrary set. By a *countable algebra* of subsets of X we mean a countable collection A of subsets of X closed under the operation of taking complements, finite intersections and hence finite unions. A *finite partition* P of X is a collection $\{S_1, \ldots, S_k\}$ of disjoint sets whose union is X. Notice that the collection of all finite unions of elements of a finite partition of X forms a countable algebra.

Given two finite partitions P_1 and P_2 of X we can form their *span*, $P_1 \vee P_2$ consiting of all intersections of one element of P_1 and one of P_2. We say P' *refines* P if every element of P is a union of elements of P'. Thus $P_1 \vee P_2$ is the smallest partition refining both P_1 and P_2.

Starting with a countable algebra $A = \{A_1, A_2, \ldots\}$, we can define a refining sequence of partitions P_0, P_1, ... in A as follows. Let $Q_i = \{A_i, A_i^c\}$ and now set $P_0 = \{X\}$ and $P_{i+1} = P_i \vee Q_{i+1}$. Notice that any set in A is a finite union of elements of some P_i in the sequence.

Let $P_i = \{S_{i,1}, S_{i,2}, \ldots, S_{i,k(i)}\}$. As each $S_{i,j}$ is a union of sets in P_{i+1}, we can associate with the sequence a directed graph. Its nodes are the sets $S_{i,j}$ and an arrow goes from $S_{i,j}$ to $S_{i+1,j'}$ if $S_{i+1,j'} \subseteq S_{i,j}$. The sequence P_1 and associated directed graph we call a refining tree of partitions.

Thus from a countable algebra we can construct a refining tree of partitions. Notice that conversely, given a refining tree of partitions the collection of all finite unions of elements of the various P_i is a countable algebra. Notice also that the tree assoicated with a countable algebra is not unique. If we re-order the elements of A in our description, the tree may change.

Example 1 Let P_i consist of intervals $[k/2^i, (k+1)/2^i)$ in $[0, 1)$.

Example 2 Let P_i consist of squares of the form $[k/2^i, (k+1)/2^i) \times [k'/2^i, (k'+1)/2^i)$ in the unit square $[0, 1) \times [0, 1)$.

2.2 Generating trees and additive set functions

We say a tree of partitions *generates* if for any $x_1, x_2 \in X$, there is a partition P_i with x_1 and x_2 in distinct elements of P_i. Both Examples 1 and 2 generate.

An *additive set function* on a tree of partitions $\{P_i\}$ is a map μ_0 from the sets of each P_i to $\mathbb{R}^+$ with

(1) $\mu_0(\emptyset) = 0$;

(2) $\mu_0(X) = 1$; and

(3) if $A = \bigcup_{j=1}^k A_j$ a disjoint finite union, where $A \in P_i$, $A_j \in P_{i+1}$, then

$$\mu_0(A) = \sum_{j=1}^{k} \mu_0(A_j). \tag{2.1}$$

Examples

(1) $\mu_0(S) =$ length of S.

(2) $\mu_0(S) =$ area of S.

We are, at this point, interested in generating trees of partitions, with a given additive set function defined on the sets of the corresponding algebra. A chain of sets $\mathscr{C} = \{c_1, c_2, \ldots\}$ in a tree of partitions is a sequence $c_i \in P_i$ where $c_{i+1} \subset c_i$. For any such chain, $\mu_0(c_i)$ is a non-increasing sequence, hence converges to some value we call $\mu(\mathscr{C})$.

Exercise 2.1 Show that $\mu(\mathscr{C}) > 0$ on at most a countable collection of chains

$$\Lambda = \{\mathscr{C}_1, \mathscr{C}_2, \ldots\} \quad \text{and} \quad \sum_{i=1}^{\infty} \mu(\mathscr{C}_i) \leq 1. \tag{2.2}$$

As we assume $\{P_i\}$ generates, for any chain $\mathscr{C}$,

$$x(\mathscr{C}) = \bigcap_{i=1}^{\infty} c_i \tag{2.3}$$

consists of either one point, or is $\varnothing$. Those points which are such intersections with $\mu(\mathscr{C}) > 0$ are called *atoms* of μ_0. We say μ_0 is *non-atomic* if no such atoms exist, i.e., for all chains $\mathscr{C}$, $\mu(\mathscr{C}) = 0$.

Exercise 2.2 If μ_0 is non-atomic then $\lim_{i\to\infty} \mu_0(c_i) = 0$, uniformly on chains $\mathscr{C}$.

Let E be the union of all sets A in any P_i with $\mu_0(A) = 0$. Let $X_0 = X\backslash E$.

Theorem 2.1 *If $\{P_i\}$ is a generating tree of partitions of X, and μ_0 is a non-atomic additive set function on it, then there is a 1–1 map*

$$\phi : X_0 \to Z \subset \mathbb{R}.$$

Z is a compact totally disconnected subset of $\mathbb{R}$ and for any set $A \in P_i$, $\overline{\phi(A)}$ is open and closed in the relative topology on Z and $\mu_0(A) = \ell(\overline{\phi(A)})$, (remember ℓ is Lebesgue measure).

Proof To each $A \in P_i$, $\mu_0(A) > 0$, we will assign a closed interval $I(A) \in \mathbb{R}^+$ as follows. For $P_0 = \{X\}$, assign an interval $I(X)$, with $\ell(I(X)) = 5/4$. Assume we have made assignments through P_k and

(1) if A, A' are disjoint in P_k so are $I(A)$, $I(A')$;

(2) if $A \subset A'$ are in different P_j's, then $I(A) \subset I(A')$; and

(3) if $A \in P_k$ then

$$\mu_0(A)(1 + \varepsilon_k) < \ell(I(A)) < \mu_0(A)(1 + 2^{-k})$$

for some $2^{-k} \geq \varepsilon_k > 0$.

To construct the intervals for P_{k+1}, simply select inside each $I(A)$, $A \in P_k$, disjoint closed intervals one for each $A' \subset A$, $A' \in P_{k+1}$, each some small fraction $\varepsilon_{k+1} \leq 2^{-k-1}$ larger than $\mu_0(A') > 0$.

For any chain $\mathscr{C} = \{c_i\}$ there corresponds a constructed chain of closed intervals $I(c_i)$ and as

$$\mu_0(c_i)(1 + 2^{-i}) > \ell(I(c_i)), \tag{2.4}$$

$\bigcap_k I(c_k)$ is a single point.

For any $x \in X_0$, let $c_k(x) \in P_k$ be that set containing x and $\phi(x)$ be the point in $\mathbb{R}^+$ corresponding to this chain. This defines ϕ. As $\mu_0(A) = 0$ implies $(A \cap X_0) = \varnothing$, $\phi(A) \cap I(A) \neq \varnothing$ for any set A. It follows that if $A \in P_{j_0}$,

$$\overline{\phi(A)} = \bigcap_{j=j_0+1}^{\infty} \left(\bigcup_{\substack{A_i \in P_j \\ A_i \subset A}} I(A_i) \right) \tag{2.5}$$

(as μ_0 is non-atomic, for $A \in P_j$, $\ell(I(A))$ goes to 0 uniformly as $j \to \infty$, and $\phi(A) \cap I(A_i) \neq \varnothing$ if $A_i \subset A$). As

$$\mu(A)(1 + \varepsilon_j) < \ell \bigcup_{\substack{A_i \in P \\ A_i \subset A}} I(A_i) < \mu(A)(1 + 2^{-j}),$$

$$\ell(\overline{\phi(A)}) = \mu_0(A). \tag{2.6}$$

Since $\overline{\phi(A)}$ and $\overline{\phi(A^c)}$ are disjoint for $A \in P_i$, these sets are both open and closed in the relative topology on $Z = \overline{\phi(X)}$.

The construction of the map ϕ is simply modelling μ_0 and X by a Cantor-like set in $\mathbb{R}$ of Lebesgue measure 1.

Exercise 2.3 State and prove a modified version of this theorem that includes the possibility of atoms.

2.3 Lebesgue spaces

We call X, with generating tree of partitions $\{P_i\}$ and non-atomic set functon μ_0, *Lebesgue* if

$$\ell(\overline{\phi(X_0)} \backslash \phi(X_0)) = 0. \tag{2.7}$$

In more technical terms Lebesgue means that

(1) for any $\varepsilon > 0$ there is a set

$$E(\varepsilon) = \sum_{i=1}^{\infty} A_i,$$

where each $A_i \in P_j$ for some j and the A_i are pairwise disjoint with

$$\sum_i \mu_0(A_i) < \varepsilon;$$

and

(2) for any chain $\mathscr{C} = \{c_1\}$ with $\bigcap c_i = \varnothing$, once i is sufficiently large, $c_i \subset E(\varepsilon)$, or

$$\textit{the empty chains have measure } 0. \tag{2.8}$$

Exercise 2.4 The definition above of a non-atomic Lebesgue space is given in two forms (2.7) and (2.8). Show that they are equivalent. Hint: first show that any union of sets in a tree can be written as a disjoint union.

Exercise 2.5 Both Examples (1) and (2) are Lebesgue.

At this point being *Lebesgue* seems to depend critically on the choice of $\{P_i\}$. We proceed to eliminate this artifact. From now on we assume X, $\{P_i\}$, μ_0 is Lebesgue, and non-atomic unless otherwise stated.

In $Z = \overline{\phi(X)}$ we have the σ-algebra of Lebesgue measurable sets $\mathscr{F}$. Let $\mathscr{A}$ be the inverse images in X of such sets, a σ-algebra in X containing all the P_i. For any $A \in P_i$, $\phi(A)$ is Lebesgue measurable as $\mathscr{F}$ is complete.

Remember (Royden 1968) $\mathscr{F}$ consists of those sets whose outer and inner measures are equal. For $A \in P_i$, $\phi(A)$ is within measure 0 of $\overline{\phi(A)}$. These closed and open sets generate the topology on Z. If we define $\mathscr{C}(S)$ to be the set of all coverings of S by disjoint unions of sets A, each an element of some P_j, and then define

$$\mu^*(S) = \inf_{C \in \mathscr{C}(S)} \sum_{A \in \mathscr{C}} \mu_0(A) = \ell^*(\phi(S)), \tag{2.9}$$

then for $S \in \mathscr{A}$,

$$\begin{aligned}\mu^*(S) &= \ell(\phi(S)) \\ &= 1 - \ell(\phi(S^c)) \\ &= 1 - \mu^*(S^c),\end{aligned}$$

and if

$$\mu^*(S) = 1 - \mu^*(S^c)$$

then

$$\ell^*(\phi(S)) = 1 - \ell^*(\phi(S^c))$$

and $S \in \mathscr{A}$. Thus

$$\mathscr{A} = \{S : \mu^*(S) = 1 - \mu^*(S^c)\} \tag{2.10}$$

and we write $\mu(S)$ for $\ell(\phi(S)) = \mu^*(S)$. The extension of μ_0 to all of $\mathscr{A}$ under the Lebesgue hypothesis is a version of the Kolmogorov extension theorem (Chung 1968). It can, of course, be done directly in terms of outer and inner measure. The value of our approach via the injection ϕ to $\mathbb{R}$ is that we can now use the very tight connection in $\mathbb{R}$ between geometry and measure.

We want to see that if one generating tree of partitions and additive set function μ_0 yields a Lebesgue space, then any other choice for a tree from $\mathscr{A}$ is also Lebesgue. If $(X, \mathscr{A}, \mu)$ has a choice for X, $\{P_i\}$, μ_0 making it Lebesgue, we call $(X, \mathscr{A}, \mu)$ a Lebesgue space.

Let $\{Q_i\}$ be a tree of finite partitions, not necessarily generating, made of sets in $\mathscr{A}$. We first create a new space $\bar{X}$ on which it does generate. We say $x_1 \sim x_2$ if for any i and $S \in Q_i$, if $x_1 \in S$ then $x_2 \in S$. This is an equivalence relation. Let $\bar{X}$ be the space of equivalence classes of $\sim$. The Q_1 can be thought of as partitions of $\bar{X}$ and on this space, $\{Q_i\}$ generates.

Theorem 2.2 *If $(X, \mathscr{A}, \mu)$ is Lebesgue and non-atomic and $\{Q_i\}$ is any tree of partitions from $\mathscr{A}$ with $\mu(S) > 0$ for all $S \in Q_i$, then $(\bar{X}, \{Q_i\}, \mu)$ is Lebesgue.*

Proof What we must show is (2.8) that given any $\varepsilon > 0$, we can find a countable disjoint collection of sets B_i, each an element of some Q_i, so that

$$\sum_i \mu(B_i) < \varepsilon$$

and for any chain $\mathscr{C} = \{c_i\}$ from the tree $\{Q_1\}$ with $\bigcap c_i = \varnothing$, $c_k \subset \bigcup B_i$, for some k. We fix ε. We will work in Z, the image space constructed in Theorem 2.1 using the Lebesgue space $(X, \{P_i\}, \mu_0)$ we know exists.

For each set $S \in Q_i$ we construct a closed subset $D(S) \subset \phi(S)$ inductively as follows. First, let $D_0 \subset \phi(X)$ be a closed subset with

$$\ell(D_0) > 1 - \varepsilon/2.$$

This exists as $\phi(X)$ is measurable.

Inside each set $\phi(S)$, $S \in Q_1$, find a closed subset $D(S)$ (perhaps $\varnothing$), where $D(S) \subset D_0 \cap \phi(S)$ with

$$\ell(D(S)) \geq \ell(D_0 \cap \phi(S))(1 - \varepsilon/4).$$

Assuming we have assigned closed sets $D(S)$ through Q_k, if $S \in Q_{k+1}$, $S \subset S' \in Q_k$, select a closed set (perhaps $\varnothing$)

$$D(S) \subset \phi(S) \cap D(S')$$

with

$$\ell(D(S)) \geq \ell(\phi(S) \cap D(S'))\left(1 - \frac{\varepsilon}{2^{k+1}}\right). \tag{2.11}$$

Let $\psi_k = \bigcup_{S \in Q_k} D(S)$, a finite disjoint union of closed sets. Clearly $\psi = \bigcap \psi_k$ is closed and contained in Z. Furthermore

$$\ell(\psi) \geq \sum_{k=1}^{\infty}\left(1 - \frac{\varepsilon}{2^k}\right) > 1 - \varepsilon. \tag{2.12}$$

Let $\mathscr{S} = \{S \in Q_1, \text{ some } i : \phi(S) \cap \psi = \varnothing\}$, a countable set. Let $\{S_i\}$ be a covering of all elements of $\mathscr{S}$ by disjoint elements of $\mathscr{S}$. Note that if $S_1, S_2 \in \mathscr{S}$ and $S_1 \cap S_2 \neq 0$ then either $S_1 \subset S_2$ or $S_2 \subset S_1$.

Now

$$\mu\left(\bigcup_{S \in \mathscr{S}} S\right) = \sum_i \mu(S_i) < 1 - \ell(\psi) < \varepsilon. \tag{2.13}$$

We want to see that these cover the empty chains. If $\mathscr{C} = \{c_i\}$ is a chain in $\{Q_1\}$ with $\bigcap c_i = \varnothing$, as $D(c_i) \subset \phi(c_i)$ and $\bigcap_i \phi(c_i) = \varnothing$ we know $\bigcap_i D(c_i) = \varnothing$. But $D(c_i)$ are closed, hence for *some* i, $D(c_i) = \varnothing$ and for this i, $\phi(c_i) \cap \psi = \varnothing$ hence $c_i \in \mathscr{S}$ and $c_i \subset \bigcap_i S_i$, and we are done. ∎

Exercise 2.6 State an prove a modified version of this theorem that includes the possibility that $\{Q_i\}$ has atoms.

Knowing that $(X, \{Q_i\}, \mu)$ is Lebesgue, we can apply Theorem 2.1 to map the equivalence classes $\bar{X}$ into a compact subset of $\mathbb{R}$ by a map ϕ. The completion with respect to μ of the σ-algebra of sets ϕ^{-1} (Lebesgue set) we write as $\bigvee_{i=1}^{\infty} Q_i$, and call such sets $\{Q_i\}$ measurable. We call the measure space $(\bar{X}, \bigvee_{i=1}^{\infty} Q_1, \mu)$ a *factor* of $(X, \mathscr{F}, \mu)$.

We next want to see that generating is all we need in order to specify the Lebesgue sets, i.e., two different generating trees of partitions yield the same σ-algebra of measurable sets.

Theorem 2.3 *If $(X, \mathscr{A}, \mu)$ is Lebesgue and non-atomic and $\{Q_i\}$ is a generating tree of partitions then $\bigvee_{i=1}^{\infty} Q_i = \mathscr{A}$.*

Proof From the previous theorem we know that there is a decreasing sequence of sets E_i, each of which is a countable disjoint union of sets from various Q_j's $\mu(E_i) \underset{i}{\to} 0$ and each E_i covers the empty $\{Q_i\}$ chains. Let $\{P_i\}$ be the partitions generating $\mathscr{A}$.

Let $H_i = P_i \vee Q_i \vee (\bigvee_{j=1}^{i} (E_j, E_j^c))$, the minimal partition for which P_i, Q_i and $E_j, j = 1, \ldots, i$ are all finite unions of elements of H_i. Delete from X all sets $S \in H_i$ with $\mu(S) = 0$, a set of total measure 0, leaving $X' \subseteq X$, a measurable subset of full measure. Now X', $\{P_i\}$, μ_0 is still Lebesgue and showing $\{Q_i\}$ generates $\mathscr{A}$ on X' implies it does on X. In this case, however, for all $S \in H_i$, $\mu(S) > 0$ and without loss of generality we write X for X'.

We know X, $\{H_i\}$, μ is Lebesgue and non-atomic. Using Theorem 2.1, if ϕ is the injection to $\mathbb{R}$ constructed using $\{H_i\}$, μ then $\overline{\phi(X)} = Z$ is a compact totally disconnected subset $Z \subseteq \mathbb{R}^+$.

For any sets $S \in H_i$, $\overline{\phi(S)}$ is a compact open subset of Z in the relative topology. Hence $E \bigcap_i \overline{\phi(E_i)}$ is a closed subset of measure 0 in Z, as E_i is a finite union of elements of H_i. As $\{Q_i\}$ generates, if $z_1, z_2 \in \phi(X)$, there are disjoint sets $S_1, S_2 \in Q_i$ for some i with

$$z_1 \in \overline{\phi(S_1)}, \quad z_2 \in \overline{\phi(S_2)}. \tag{2.14}$$

Note: S_1 and S_2 are each finite disjoint unions of elements of H_i, hence $\overline{\phi(S_1)}$, $\overline{\phi(S_2)}$ are disjoint.

Notice $E^c \cap Z \subset \text{Range}(\phi)$. If $z_1 \in E$ and $z_2 \in E^c \cap Z$ we claim there must be disjoint $S_1, S_2 \in Q_i$ for some i with

$$z_1 \in \overline{\phi(S_1)}, \quad z_2 \in \overline{\phi(S_2)}. \tag{2.15}$$

Otherwise z_1 and z_2 are always in the same $\overline{\phi(S_i)}$, $S_i \in Q_i$. As $z_1 \in E$, this forces $z_2 \in E$, as its Q_i chain is always contained in $\overline{\phi(E_i)}$.

Let C be a closed subset of Z. We want to show $\phi^{-1}(C) \in \mathscr{B} = \bigvee_{i=1}^{\infty} Q_i$. By (2.14) and (2.15), for any $z_1 \in C$, $z_2 \in C^c \cap E^c \cap Z$ there are disjoint sets

$S_1(z_1, z_2)$, $S_2(z_1, z_2) \in Q_i$ for some i and

$$z_1 \in \overline{\phi(S_1(z_1, z_2))}, \quad z_2 \in \overline{\phi(S_2(z_1, z_2))}.$$

Fix z_2 and consider $\{\phi(S_1(z_1, z_2)) | z_1 \in C'\}$ an open cover of C. As C is compact, there is a finite subcover. Let $\overline{\phi(S(z_2))}$ be the union of elements in this finite collection of sets, again an open compact subset of Z and $z_2 \in \overline{\phi(S(z_2))}$. Now $S(z_2)$ is a finite union of elements of some Q_i.

Let

$$C' = \bigcap_{z_2 \in E^c \cap C^c} \phi(S(z_2)), \tag{2.16}$$

at most a countable intersection as $S(z_2)$ is an element of a countable algebra of sets. Hence $\phi^{-1}(C') \in \mathscr{B}$. But $C \subset C' \subset C \cup E$ and as $\mathscr{B}$ is complete $\phi^{-1}(C) \in \mathscr{B}$ and we are done. ■

Note: For our purposes, then, a Lebesgue probability space is just a measurable subset of measure 1 of a compact totally disconnected subset of $\mathbb{R}$.

This theorem gives us the following corollary.

Corollary 2.4 *If $(X, \mathscr{A}, \mu)$ and $(X, \mathscr{A}', \mu')$ are both Lebesgue probability spaces, and $\phi: X \to X'$ is* 1–1, *onto and measurable, and furthermore*

$$\mu(\phi^{-1}(A)) = 0 \quad \textit{only if } \mu'(A) = 0$$

(called non-singularity), then ϕ^{-1} is also measurable.

Proof Let $\{P_i\}$ be a generating tree of measurable partitions for $(X', \mathscr{A}', \mu')$.

As ϕ is 1–1, onto and non-singular, $\{Q_i\} = \{\phi^{-1}(P_i)\}$ is a generating tree of partitions in $\mathscr{A}$. Hence

$$\phi^{-1}(\mathscr{A}) = \phi^{-1}\left(\bigvee_{i=0}^{\infty} P_i\right) = \bigvee_{i=0}^{\infty} \phi^{-1}(P_i) = \bigvee_{i=0}^{\infty} Q_i = \mathscr{A}'.$$

Thus for $S \in \mathscr{A}'$, $\phi(S) \in \mathscr{A}$ and ϕ^{-1} is measurable. ■

We will indicate a Lebesgue space by a triple $(X, \mathscr{F}, \mu)$ where $\mathscr{F}$ is the σ-algebra of measurable sets and μ is the measure. It will be useful, at times, to select a particular generating tree of partitions and assume they are open compact subsets. In this sense the significance of our work in the previous arguments is not so much to show the existence of one identification of our space with a subset of $\mathbb{R}^+$, but rather to understand their multitude.

Exercise 2.7 If $(X, \mathscr{F}, \mu)$ is a non-atomic Lebesgue space, then there is a subset $X_0 \subseteq X$, with X_0^c a countable set and a 1–1 measure-preserving map $\phi: X_0 \to (0, 1)$.

Exercise 2.8 Modify the previous exercise to include the possibility of atoms.

Exercise 2.9 Returning to Example 3 of Chapter 1, show that μ_p, given as a finitely additive set function on cylinder sets, makes $\sum_M$ a Lebesgue space.

Exercise 2.10 Returning to Example 4 of Chapter1, show that Haar measure on a compact abelian monothetic metric group makes it a Lebesgue space.

Having seen what a Lebesgue probability space is, it behooves us to consider what can fail in the arguments we have given if less is assumed. On a space X let $\{P_i\}$ be a generating tree of partitions, and μ_0 an additive set function on the elements of the partition.

In Theorem 2.1 we saw that this is enough to construct a 1–1 map ϕ from X to some subset $\text{Range}(\phi) \subset \mathbb{R}$. We know

$$\ell(\overline{\text{Range}(\phi)}) = 1.$$

As ϕ is 1–1, we may assume $X = \text{Range}(\phi) \subset \mathbb{R}$. We know μ_0 can be extended to a measure μ making $(X, \mathscr{F}, \mu)$ Lebesgue if

$$\ell(\overline{X} \backslash X) = 0. \tag{2.17}$$

This can fail to occur at two levels, first μ_0 might not be extenable to a measure μ, and second $(X, \mathscr{F}, \mu)$ might not be Lebesgue. Here are examples of both possibilities.

Let $X \subset [0, 1]$ be a measurable subset so that on any interval I,

$$\ell(I) > \ell(I \cap X) > 0.$$

Let P_i be the partition of X into intervals of width 2^{-i}, and for $S_{i,j} \in P_i$,

$$\mu_0(S_{i,j}) = 2^{-i}.$$

In this case the additive set function μ_0 cannot be extended to a measure, as X can be covered by a countable union of dyadic intervals, the sum of whose measures is strictly less than 1.

For a second example, let $X \subset [0, 1]$ be a non-measurable subset of outer measure 1. Again let P_i be the partition of X into intervals of width 2^{-i} and set $\mu(S) = \ell(A)$, where A is Lebesgue measurable and $S = A \cap X$. That X is of full outer measure implies this definition is independent of the choice of A. The measure space is, though, not Lebesgue as

$$\ell^*(\overline{\text{Range}(\phi)} \backslash \text{Range}(\phi)) = \ell^*(X^c \cap [0, 1] > 0, \tag{2.18}$$

(the map ϕ will, in fact, be 1–1 and measure-preserving on $[0, 1]$). This example is the general form of a separable probability space. A probability space could also fail to be Lebesgue if no tree of partitions could generate, i.e., if the cardinality of X were larger than c.

We also have seen in Theorem 2.3 that in a Lebesgue space for a tree of partitions to separate points is equivalent to generating the measure algebra. The following example shows that this also can fail in the non-Lebesgue case.

Exercise 2.11 Let $A \subseteq [0, \frac{1}{2})$ be a subset with $\ell^*(A) = \frac{1}{2}$ and $\ell^*(A^c \cap [0, \frac{1}{2})) = \frac{1}{2}$. Let $X = A \cup (A^c + \frac{1}{2}) \subseteq [0, 1]$.

Define $\mathscr{F} = \{X \cap E : E$ is Lebesgue measurable$\}$ and $\mu(X \cap E) = \ell(E)$ as in the above example

(1) show μ is well defined;

(2) exhibit a tree of partitions $\{P_i\} \subseteq \mathscr{F}$ that separates points of X but does *not* generate the full measure algebra $\mathscr{F}$.

At the other extreme from these examples is a case which cannot fail to be Lebesgue. This is the case when our generating tree of partitions has no empty chains. For example the Cantor-like set Z constructed in Theorem 2.1 has this property as do mixing Markov chains, for the given tree of partitions. In such cases all additives set functions on the tree extend to measures. One can topologize such a space, taking as a base for the open sets the sets in the tree. The topology is totally disconnected, metrizable and, as there are no empty chains, compact. The Borel measures on this space are, by the Riesz representation theorem, the dual of the continuous functions. In this particular case the Riesz theorem is easily proven, as the characteristic functions of sets in the tree are continuous. Borel probability measures on this space are exactly the ones which come from additive set functions on the tree and are the same as positive normed linear functionals on the continuous functions. This space of Borel probability measures is itself a compact metric space, as the dual of the continuous functions. We will see much more of this in Chapter 6.

2.4 A martingale theorem and conditional expectation

As an application of our Lebesgue space development we will prove a martingale theorem. The result is quite standard and our proof is not particularly special. It is though, via a covering argument and the Vitali lemma. Such covering arguments are the core of our approach to many later results and we wish to stress the analogy and so include the complete argument.

Lemma 2.5 (Vitali covering lemma) *Let $S \subset [0, 1]$ be Lebesgue measurable and suppose that for each $x \in S$ there is given a nested sequence of intervals $I_i(x)$ which intersect to x. Then there is a countable disjoint collection $I_1, I_2, \ldots$ where $I_k = I_{i(k)}(x(k))$ for some $i(k)$, $x(k) \in S$ which covers almost all of S, i.e.,*

$$\ell\left(S \Big\backslash \bigcup_{k=1}^{\infty} I_k\right) = 0. \tag{2.19}$$

Proof Select the $\hat{I}_k$ inductively with length $\hat{I}_{k+1} > \frac{1}{2}\sup\{\text{length } I_j(x) \mid I_j(x)$ disjoint from $\hat{I}_1, \hat{I}_2, \ldots, \hat{I}_k\}$, with $\hat{I}_{k+1}$ disjoint from $\hat{I}_1, \ldots, \hat{I}_k$. Either this process terminates after finitely many steps, in which case we are finished or we get a sequence of disjoint intervals $\hat{I}_1, \hat{I}_2, \ldots$.

The length of $\hat{I}_i \to 0$ as $i \to \infty$. In fact, the sum of the lengths of all the $\hat{I}_i$ which are selected is bounded by 1. Let I_i' be $\hat{I}_i$ expanded symmetrically by a fraction ε of its length. So the length of $I_i' = (1 + \varepsilon) \times$ length $\hat{I}_i$. Let $\bar{I}_i$ be $\hat{I}_i$ expanded (symmetrically) 5 times its length.

Claim. For all $t \geq 2$, all $\varepsilon > 0$

$$S \subset \bigcup_{i=1}^{t} I_i' \cup \bigcup_{i=t+1}^{\infty} \bar{I}_i. \tag{2.20}$$

Otherwise there is some $x \in S$ not in the right hand side. So $x \notin \bigcup_{i=1}^{t} I_i'$, and there is some $I_j(x)$ disjoint from $\hat{I}_1, \ldots, \hat{I}_t$.

If $I_j(x)$ is also disjoint from $\hat{I}_{t+1}, \hat{I}_{t+2}, \ldots$ then we get a contradiction to the selection of the $\hat{I}_i$ (length $I_j(x) >$ twice the length $\hat{I}_t$ for large enough t). If $\hat{I}_s$ is the first of $\hat{I}_{t+1}, \hat{I}_{t+2}, \ldots$ which intersects $I_j(x)$, then $x \notin \bar{I}_s$ and so length $(I_j(x)) > 2$ length $(\hat{I}_s)$ and this is another contradiction to the selection of $\hat{I}_s$.

Therefore (2.20) is true so

$$\ell\left(S \Big\backslash \bigcup_{i=1}^{t} \hat{I}_i\right) \leq \ell\left(\bigcup_{i=1}^{t} [I_i' \backslash \hat{I}_i]\right) + \ell\left(\bigcup_{t+1}^{\infty} \bar{I}_i\right) < \varepsilon + \varepsilon$$

for large t. ∎

Let $\{Q_1\}$ be a tree of partitions for X, not necessarily generating. Recall we can construct a space X' of equivalence classes on which $\{Q_i\}$ generates and is Lebesgue. The purpose of the martingale theorem we now want to prove is to project $L^1(\mu)$ on X to $L^1(\mu')$ on X'. Such a projection is called a *conditonal expectation.* Here are some simple examples of such projections.

Example 1 Suppose $(X, \mathscr{F}, \mu)$ is the unit interval with Lebesgue measure, and $B_0 = \{[0, \frac{1}{2}), [\frac{1}{2}, 1), [0, 1), \varnothing\}$. The factor space X' consists of two points, call them $\{R, B\}$. If $f \in L^1(\mu)$ we define

$$(pf)(x') = \begin{cases} 2\displaystyle\int_{[0,\frac{1}{2}]} f\,\mathrm{d}\mu & \text{if } x' = R \\ 2\displaystyle\int_{[\frac{1}{2},1]} f\,\mathrm{d}\mu & \text{if } x' = B. \end{cases}$$

Then for any $A \in X'$ (i.e., $A \in \{R, B\}$) we have

$$\int_A p(f)\,\mathrm{d}\mu' = \int_{p^{-1}(A)} f\,\mathrm{d}\mu.$$

Example 2 Let $\mathscr{F}'$ be 'doubled sets' in $[0,1]$, i.e., if $x \in B$, $B \in \mathscr{F}'$, then $x \pm \frac{1}{2} \in B$ for the appropriate $\pm$ sign. The equivalence classes of $[0,1] \bmod \mathscr{F}'$ consist of pairs of points $\{x, y\}$ where $|x - y| = \frac{1}{2}$.

If $f \in L^1([0,1])$ let $(pf)(\{x,y\}) = \frac{1}{2}(f(x) + f(y))$ and for $A \in \mathscr{F}'$

$$\int_A p(f)\,\mathrm{d}\mu' = \int_{p^{-1}(A)} f\,\mathrm{d}\mu.$$

So $p : L^1(\mu) \to L^1(\mu')$ isometrically.

Let $\{Q_i\}$ be a tree of partitions in a Lebesgue probability space $(X, \mathscr{F}, \mu)$. Let $f \in L^1_+(\mu)$, the positive integrable functions. Consider the space $(X', \mathscr{F}', \mu')$ which $\{Q_i\}$ generates. We want to project f to an $f' \in L^1(\mu')$ so that for any $S \in \mathscr{F}'$,

$$\int_S f'\,\mathrm{d}\mu' = \int_{p^{-1}(S)} f\,\mathrm{d}\mu. \tag{2.21}$$

We call f' the conditional expectation of f given $\mathscr{F}'$, and will write it $E(f|\mathscr{F}')$.

For a finite partition Q we can easily define the conditional expectation of f given Q as

$$f_Q(x) = \frac{1}{\mu(S)} \int_S f\,\mathrm{d}\mu, \quad \text{where } x \in S \in Q, \tag{2.22}$$

If $\{Q_i\}$ is a tree of partitions, we see each f_{Q_i} is a simple function constant on each $S \in Q_i$. Further, for $S \in Q_i$ and $j \geq i$

$$\int_S f_{Q_i}\,\mathrm{d}\mu' = \int_S f_{Q_j}\,\mathrm{d}\mu'. \tag{2.23}$$

In fact, we need not assume f_{Q_i} actually arises from some original $f \in L^1_+$; only these last conditions (2.23) need be assumed to get the result we want.

We call $\{f_i\}$, $\{Q_i\}$ a *martingale* if $\{Q_i\}$ is a tree of partitions, f_i is constant on sets $S \in Q_i$ and for $S \in Q_i$ and $j \geq i$

$$\int_S f_i\,\mathrm{d}\mu = \int_S f_j\,\mathrm{d}\mu. \tag{2.24}$$

This notion of a martingale is much stronger than is standard, and so our martingale convergence theorem is a much weakened version of Doob's martingale theorem (Chung 1968).

Theorem 2.6 *If $\{f_i\}$, $\{Q_i\}$ is a martingale, $f_i \geq 0$, then f_i converges a.e. to a function $f \in L^1(\mu)$, and for any $\{Q_i\}$ measurable set S*

$$\underline{\lim} \int_S f_i \geq \int_S f. \tag{2.25}$$

Notice, as each f_i is Q_i measurable, we can assume without loss of generality $\{Q_i\}$ generates.

Proof First, we show that $S = \{x : f_i(x) \to \infty\}$ has measure 0. S is measurable so for any $\varepsilon > 0$, we can find a finite disjoint union of sets $S_j \in Q_i$ for some i with $\mu(\bigcup_j S_j \Delta S) < \varepsilon$.

As $\bigcup_j S_j$ is a finite union, once i is large enough

$$\int_{\bigcup_j S_j} f_i \, \mathrm{d}\mu$$

is a constant in i. But now

$$\int_{S \cap \bigcup_j S_j} f_i \leq \int_{\bigcup_j S_j} f_i < \infty$$

and we conclude $\mu(S \cap \bigcup_j S_j) = 0$ and $\mu(S) < \varepsilon$. Thus $\mu(S) = 0$.

Now to show f_i converges pointwise a.e. to a limit $\bar{f}$.

It will be convenient to transport our construction to $\mathbb{R}$, using the $\{Q_i\}$ trees, so that each set $\overline{\phi S}$, $S \in Q_i$, is an open compact set, with $\phi(S) \subset I(S)$ an interval of length less than $\mu(S)(1 + 2^{-i})$.

Since $\{Q_i\}$ may contain atoms the tree of intervals may have branches which descend to a set of positive measure. Such a set will be a closed interval. Thus ϕ will not be defined pointwise on atoms of μ, but will map the atom to the whole interval as a set map. (If you did not solve Exercise 2.3, take this as a hint and do so now.)

We may define $\bar{f}_i$ on $\overline{\phi(X)}$, by transporting to each $I(S)$ the value of f_i on $S \in Q_i$. Clearly if $\bar{f}_i$ converges a.e. on $\phi(X)$ then f_i does on X. *Note*: If $C_1 \supset C_2 \supset \cdots$ is a branch of $\{Q_i\}$ with $\lim_{i \to \infty} \mu(C_1) > 0$, i.e., an atom, then clearly $f_i(x)$ converges as $i \to \infty$ for all $x \in \bigcap C_i$ as

$$0 \leq f_i(x)\mu(C_i) \leq f_{i-1}(x)\mu(C_{i-1}), \tag{2.26}$$

i.e., $f_i(x)$ is nearly monotone decreasing. Thus we need only work on the non-atomic part.

Pick $a > \varepsilon > 0$ and let

$$S_{a,\varepsilon} = \{z \in \phi(X) : \bar{f}_i(z) > a + \varepsilon \text{ infinitely often and } \bar{f}_i(z) < a - \varepsilon \text{ infinitely often}\}.$$

$f_i(x)$ converges a.e. iff $\ell(S_{a,\varepsilon}) = 0$ for all a, ε. Notice also that if $\bar{f}_i(z)$ converges then $z \notin S_{a,\varepsilon}$ for any a, ε.

By our note if $z \in S_{a,\varepsilon}$ then $\phi^{-1}(z)$ is not in an atom of μ. Thus, for every $z \in S_{a,\varepsilon}$ we can select intervals $I_i(z)$ from the construction of ϕ such that $I_{i+1}(z) \subseteq I_i(z)$, $\ell(I_i(z)) \to 0$ and for each i, z if $j \geq j(i, z)$ then $I_i(z)$ corresponds to some $Q_{j(i,z)}$ so that

(1) if i is even

$$\int_{I_i(z)\cap\overline{\phi(x)}} \bar{f}_{j(i,z)}\,\mathrm{d}\ell = \int_{I_i(z)\cap\overline{\phi(x)}} \bar{f}_j\,\mathrm{d}\ell > \left(a+\frac{\varepsilon}{2}\right)\ell(I_i(z)) \tag{2.27}$$

and

(2) if i is odd then

$$\int_{I_i(z)\cap\overline{\phi(x)}} \bar{f}_{j(i,z)}\,\mathrm{d}\ell = \int_{I_i(z)\cap\overline{\phi(x)}} \bar{f}_j\,\mathrm{d}\ell < \left(a-\frac{\varepsilon}{2}\right)\ell(I_i(z)). \tag{2.28}$$

We accomplish this by alternately selecting intervals from the two available infinite sequences for $z \in S_{a,\varepsilon}$.

We successively apply the Vitali lemma first to select a cover of $S_{a,\varepsilon}$ a.s. by a disjoint collection of intervals of even index. Call the union of these intervals U_1. We restrict the odd intervals to only those contained in U_1, and select a cover a.s. of $U_1 \cap S_{a,\varepsilon}$ by disjoint intervals of odd index. Call the union of these intervals D_2. Restrict the even intervals to only those contained in D_2 to select a cover for $D_2 \cap S_{a,\varepsilon}$ a.s. by disjoint even intervals. Call the union of these intervals U_2. Continue *ad* infinitum to build countable disjoint unions of alternately even and odd index

$$U_1 \supseteq D_2 \supseteq U_2 \supseteq D_3 \ldots$$

and $\bigcap_i D_i = \bigcap_i U_i$ contains $S_{a,\varepsilon}$ a.s. Thus

$$\ell\left(\bigcap_i D_i\right) = \ell\left(\bigcap_i U_i\right) \geq \ell(S_{a,\varepsilon}). \tag{2.29}$$

Each U_i and D_i is a countable union of intervals $I_j(z)$. We can inductively define a sequence of sets $\bar{U}_i \subseteq U_i$, $\bar{D}_i \subseteq U_i$, $\bar{U}_1 \supseteq \bar{D}_2 \supseteq \bar{U}_2 \ldots$ where each is a *finite* disjoint union of such $I_j(x)$, j even for U, odd for D and we still have

$$\ell\left(\bigcap_i \overline{D_i}\right) = \ell\left(\bigcap_i \overline{U_i}\right) \geq \ell(S_{a,\varepsilon})/2. \tag{2.30}$$

As $\overline{U_i}$ is a finite disjoint union of intervals $I_j(z)$, j even

$$\lim_{j\to\infty} \int_{\overline{U}_i} \bar{f}_j\,\mathrm{d}\ell \geq \left(a+\frac{\varepsilon}{2}\right)\ell(\overline{U}_i)$$

and as $\overline{D_i}$ is a finite disjoint union of intervals $I_j(z)$, j odd

$$\lim_{j\to\infty} \int_{\overline{D}_i} \bar{f}_j\,\mathrm{d}\ell \leq \left(a+\frac{\varepsilon}{2}\right)\ell(\overline{D}_i). \tag{2.31}$$

Now $\bar{U}_i \subseteq \bar{D}_i$ so

$$\lim_{j\to\infty} \int_{\overline{U}_i} \bar{f}_j\,\mathrm{d}\ell \geq \left(a+\frac{\varepsilon}{2}\right)\ell(\bar{D}_i) \tag{2.32}$$

hence $(a + \varepsilon/2)\ell(\bar{U}_i) \leq (a - \varepsilon/2)\ell(\bar{D}_i)$. Taking limits over i, $\ell(\bar{U}_i)$ and $\ell(\bar{D}_i)$ converge to the same value, which must be 0. Thus $\ell(S_{a,\varepsilon}) = 0$.

This gives us pointwise convergence of $\bar{f}_i$ a.e. to a function $\bar{f}$. Hence on X, f_i converges pointwise a.e. to a function f. For any $\bigvee_{i=1}^{\infty} Q_i$ measurable set

$$\underline{\lim} \int_{\phi(S)} \bar{f}_i \geq \int_{\phi(S)} \bar{f}$$

by Fatou's lemma. Hence

$$\underline{\lim} \int_S f_i \geq \int_S f. \qquad \blacksquare$$

We will usually want more than pointwise convergence. The easiest condition to use that implies L^1 convergence also, and in fact the most general is *uniform integrability*. Let $S(B,i) = \{x : |f_i(x)| \geq B\}$. If f_i is integrable, then

$$\lim_{B\to\infty} \int_{S(B,i)} |f_i|\, \mathrm{d}\mu = 0.$$

We say the sequence f_i is *uniformly integrable* if this limit is uniform in i.

Theorem 2.7 *If $f_i \geq 0$ and converge pointwise a.e. to f on the Lebesgue probability space $(X, \mathscr{F}, \mu)$, then $f_i \to f$ in L^1 iff the f_i are uniformly integrable.*

Proof The only if direction we leave as an exercise for the reader, as the if direction is all we ever use. By Fatou's lemma, $f \in L^1$. Thus given ε, there is a δ so that if $\mu(E) < \delta$ then

$$\int_E f \mathrm{d}\mu < \frac{\varepsilon}{5}.$$

As the f_i are uuniformly integrable. $\mu(S(B,i)) \leq 1/B \int_{S(B,i)} |f_i|\, \mathrm{d}\mu \to 0$ in B, uniformly in i. Select B so large that for all i, $\mu(S(B,i)) < \delta$, and

$$\int_{S(B,i)} |f_i|\, \mathrm{d}\mu \leq \frac{\varepsilon}{5}.$$

Select i so large that

$$\mu\left(\left\{x : |f_i(x) - f(x)| > \frac{\varepsilon}{5}\right\}\right) < \max\left(\frac{\varepsilon}{5B}, \delta\right).$$

For i this large,

$$\int |f_i - f|\, \mathrm{d}\mu \leq \int_{\{x:|f_i(x)-f(x)|\leq\varepsilon/5\}} |f_i - f|\, \mathrm{d}\mu$$
$$+ \int_{\{x:|f_i(x)-f(x)|>\varepsilon/5\}\cap S(B,i)^c} |f_i - f|\, \mathrm{d}\mu + \int_{S(B,i)} |f_i - f|\, \mathrm{d}\mu$$

$$\leq \frac{\varepsilon}{5} + \int_{\{x:|f_i(x)-f(x)|>\varepsilon/5\}\cap S(B,i)^c} |f_i - f|\,\mathrm{d}\mu + 2\varepsilon/5$$

$$\leq \frac{3\varepsilon}{5} + B\mu\left(\left\{x : |f_i(x) - f(x)| > \frac{\varepsilon}{5}\right\}\right) \leq \varepsilon.$$

Letting $\varepsilon \to 0$ we get the result. ∎

Corollary 2.8 *Let $f \in L^1_+$ and $\{Q_i\}$ be a tree of partitions. If*

$$f_i(x) = \frac{1}{\mu(S)} \int_S f\,\mathrm{d}\mu$$

for $x \in S \in Q_i$, then $\{f_i\}$, $\{Q_i\}$ is a positive martingale and f_i converge pointwise and in L^1.

Proof That $\{f_i\}$, $\{Q_i\}$ is a positive martingale is easy. All we need check is that the f_i are uniformly integrable. Let $S(B) = \{x : \sup f_i(x) \geq B\}$. Certainly $\mu(S(B)) \to 0$ in B as the f_i converge a.e., and for all i, $S(B, i) \subset S(B)$.

Now $S(B)$ is a countable uunion of disjoint sets $I_i \in Q_{j(i)}$, where $x \in I_i$ iff $f_{j(i)}(x) > B$, and $j(i)$ is the first such

$$f_{j(i)}(x) = \frac{1}{\mu(I_i)} \int_{I_i} f\,\mathrm{d}\mu.$$

Thus for $k \leq j(i)$,

$$\int_{I_i} f_k\,\mathrm{d}\mu \leq B\mu(I_i) < \int_{I_i} f_{j(i)}\,\mathrm{d}\mu = \int_{I_i} f\,\mathrm{d}\mu \tag{2.33}$$

and for $k > j(i)$

$$\int_{I_i} f_k\,\mathrm{d}\mu = \int_{I_i} f_{j(i)}\,\mathrm{d}\mu = \int_{I_i} f\,\mathrm{d}\mu. \tag{2.34}$$

Thus for all i

$$\int_{S(B)} f_i\,\mathrm{d}\mu \leq \int_{S(B)} f\,\mathrm{d}\mu.$$

Thus, as $B \to \infty$,

$$\int_{S(B)} f_i\,\mathrm{d}\mu \to 0$$

uniformly in i. ∎

In this example, the limit function is called the conditional expectation of f given $\bigvee_{i=1}^{\infty} Q_i$. We extend the conditional expectation to all of L^1 by setting $E(f|a) = E(f^+|a) - E(f^-|a)$. The first exercise below shows that the conditional expectation is uniquely defined.

Exercise 2.12 Let $a \subset \mathscr{F}$ be a sub-σ-algebra of a Lebesgue probability space.

(a) Show that there is a tree of partitions $\{Q_i\}$ with $a = \bigvee_{i=1}^{\infty} Q_i$, and set $E(f|a) = E(f|\bigvee_{i=1}^{\infty} Q_i)$, an a-measurable function.

(b) Show that for any set $A \in a$,

$$\int_A E(f|a) = \int_A f.$$

(c) Show that if f_1 and f_2 are both a-measurable, and for any $A \in a$, $\int_A f_1 = \int_A f_2$ then $f_1 = f_2$ a.e., and hence $E(f|a)$ is uniquely defined, independent of the $\{Q_i\}$ tree.

Exercise 2.13 Show that the construction of a conditional expectation does not require the Lebesgue property, but only separability (i.e., the existence of a generating tree of partitions).
Hint: in this case, we know X can be regarded as a subset of $(0, 1]$ of full outer measure and $\mathscr{F}$ the Lebesgue sets restricted to X.

Exercise 2.14 Show that if $g \in L^1$ is a measurable and $f \in L^1$ and $f \circ g \in L^1$ then $E(f \cdot g|a) = E(f|a) \cdot g$.

Exercise 2.15 Prove the Radon–Nikodym theorem, that if ω is another positive measure on $(\Omega, \mathscr{F}, \mu)$, a Lebesgue probability space, and $\mu(E) = 0$ implies $\omega(E) = 0$, then there is an L^1 function $f: \Omega \to \mathbb{R}^+$ and

$$\omega(S) = \int_S f \, \mathrm{d}\mu.$$

2.5 More about generating trees and dynamical systems

We saw in Theorem 2.3 that if $\{Q_i\}$ is a generating tree of partitions in a Lebesgue space, then $\bigvee_{i=1}^{\infty} Q_i$ was the full algebra, i.e., the tree also *generated* the full σ-algebra. The converse of this is also almost true.

Theorem 2.9 *If $(X, \mathscr{F}, \mu)$ is a Lebesgue space, and $\{Q_i\}$ is a tree of partitions with $\bigvee_{i=1}^{\infty} Q_i = \mathscr{F}$, then there is a subset $X_0 \subseteq X$ of full measure, and $\{Q_i\}$ generates on X_0, i.e., separates points.*

Proof Let $\{A_i\}$ be a countable collection of measurable sets which separate points in X. By Corollary 2.8 and Exercise 2.12,

$$f_{i,j}(x) = \frac{\mu(A_i \cap S)}{\mu(S)}$$

where $x \in S \in Q_j$, converges a.e. to χ_{A_i}. Let

$$X_0 = \{x \in X : f_{i,j}(x) \to \chi_{A_i}(x) \text{ for all } i\}.$$

If $x_1, x_2 \in X_0$, then as $\chi_{A_i}(x_1) \neq \chi_{A_i}(x_2)$ for some i, x_1 and x_2 are separated by the $\{Q_i\}$ tree. ■

Thus in a Lebesgue space we have two equivalent notions of a generating tree, that of generating the measure algebra up to null sets, and that of separating points off a null set.

In light of this, we modify slightly our definition of a generating partition. If no measure is specified, or many are to be considered, then a generating tree must separate points. But if some particular measure is specified, then it need only separate points off a null set to be called a generator. As our spaces will never be of cardinality beyond the continuum, any tree which separates points off a null set can be modified by null sets to actually separate points. This is consistent with our definition of isomorphism, as we now can say if a tree generates on one space, then it generates on any isomorphic space.

In situations where only one measure is being considered, no subtlety will arise in ignoring null sets in this manner. Later though, especially in Chapters 6 and 7 when we must consider spaces of measures, it will be necessary to specify a tree which generates by separating points, and not allow null sets to enter the arguments unless they are universally null, i.e., of measure 0 for all measures considered.

A *dynamical system* for us is a Lebesgue space $(X, \mathscr{F}, \mu)$ and a measurable measure-preserving bijection T from X to itself. Notice that given any finite partition P we can construct a natural tree of partitions $\{\bigvee_{i=-n}^{n} T^{-i}(P)\}$ which generates the minimal T-invariant sub-σ-algebra $\bigvee_{i=-\infty}^{\infty} T^{-i}(P)$, containing P. If this σ-algebra is all of $\mathscr{F}$, we call P a *generator* or *generating partition* for the dynamical system. Later, especially in Chapters 5 and 7 we will investigate this concept more deeply. If we are given a tree of partitions there is also a natural way to build a new tree that generates the minimal T-invariant sub-σ-algebra containing the given one. We inductively define the elements of the tree so that each new level contains both forward and backward images of the previous level of the tree as well as the next level of the original tree. In such a tree the forward and backward images of any level are finite unions of elements of the next level. Hence the T-invariance of the measure can be directly read off from the measures of sets in the tree.

3 Ergodic theorems and ergodic decomposition

A fundamental tenet of naïve probability is that the probability of an event is the average rate of occurrence of that event over time, e.g., the probability of a coin landing heads is the average number of heads achieved through many tosses. Ergodic theorems are perhaps the most natural formalization of this idea. If $(X, \mathscr{F}, \mu)$ is a Lebesgue probability space and T a bimeasurable measure-preserving map of X to itself, then T can be viewed as the movement from one successive outcome of a random series of events to the next. A measurable function f can be viewed as some particular measurement on the outcome of the event. The average value of this measurement over a span of n time intervals is, then, the Cesáro average

$$A_n(f, x) = \frac{1}{n} \sum_{i=0}^{n-1} f(T^i(x)). \tag{3.1}$$

(When regarding this as a function, we write it $A_n(f)$).

That these Cesáro averages converge in various senses under various hypotheses on f form the formal content of the first half of this chapter. This will put the naïve probabilistic notion of average rate of occurrence on firm ground, and will also provide one of the fundamental tools of ergodic theory.

We first prove perhaps the simplest of the ergodic theorems.

3.1 Von Neumann's L^2-ergodic theorem

Theorem 3.1 *Let T be an invertible, bimeasurable, measure-preserving map of the Lebesgue probability space $(X, \mathscr{F}, \mu)$. If $f \in L^2(\mu)$, then $A_n(f)$ converges in $L^2(\mu)$ to a function $\bar{f}$ with*

$$\bar{f}T(x) = \bar{f}(x) \quad \textit{for a.e. } x \in X.$$

Proof Let

$$U_T(f)(x) = f(T(x)) \tag{3.2}$$

be the unitary operator on $L^2(\mu)$ induced by T.

First suppose $f \in \text{Range}(U_T - \text{id})$, i.e. $f = U_T(g) - g$ for some $g \in L^2(\mu)$. Now

$$A_n(f) = \frac{1}{n}(U_T^n(g) - g)$$

so

$$\|A_n(f)\|^2 = \frac{1}{n}\|U_T^n(g) - g\|_2 \leq \frac{2}{n}\|g\|_2$$

which goes to 0 in n.

Notice that if $\|f - g\|_2 < \varepsilon$, then $\|A_n(f) - A_n(g)\|_2 < \varepsilon$ for all n. We conclude then that if

$$f \in \overline{\text{Range}(U_T - \text{id})},$$

then $A_n(f) \to 0$ in $L^2(\mu)$.

Now suppose

$$f \in \overline{\text{Range}(U_T - \text{id})}^{\perp}.$$

(i.e., $\langle f, U_T(g) - g\rangle = 0$ for all $g \in L^2(\mu)$).

Then

$$\langle f, U_T(g)\rangle = \langle f, g\rangle \quad \text{for all } g \in L^2(\mu),$$

or

$$\langle U_T^*(f), g\rangle = \langle f, g\rangle \quad \text{for all } g \in L^2(\mu).$$

(i.e., $U_T^*(f) = f$ a.e.).

What is the adjoint $U_T^*(f)$?

$$\langle f, U_T(g)\rangle = \int f(x)g(T(x))\,\mathrm{d}\mu = \int f(T^{-1}(x))g(x)\,\mathrm{d}\mu$$

as T is measure-preserving and invertible.

Hence

$$U_T^*(f) = f(T^{-1}(x)) \quad \text{a.e.}$$

i.e.,

$$U_T^* = U_{T^{-1}}.$$

Thus

$$f(T^{-1}(x)) = f(x) \quad \text{a.e.}$$
$$f(x) = f(T(x)) \quad \text{a.e.}$$

so

$$A_n(f) = f \quad \text{a.e.}$$

This completes the proof. Notice we obtain more, having identified $\bar{f}$ as the projection of f on the subspace of T-invariant L^2-functions. ∎

This is a purely operator theoretic argument. The study of T via the unitary operator U_T leads to the ergodic theory of linear operators, a bit outside our focus here. We will see later that when U_T has pure point spectrum, it determines T completely, but for more general spectral types U_T gives only weak information about T.

3.2 Two proofs of Birkhoff's ergodic theorem

We now prove the much deeper Birkhoff pointwise ergodic theorem, that the Cesáro averages actually converge pointwise almost everywhere. We well, in fact, give two proofs of this result. The first is the semi-classical Garsia–Halmos argument, the second a far more technical backward Vitali argument following ideas of Ornstein and Weiss (1983).

Both arguments have their special strengths. The Halmos argument requires less of the measure space, but the Ornstein–Shields argument generalizes to larger group actions and other pointwise convergence theorems.

For the Halmos result, we first prove the following.

Lemma 3.2 (Maximal ergodic theorem). *Let T be a measure-preserving transformation on the σ-finite measure space $(X, \mathscr{F}, \mu)$. Suppose $f \in L^1(\mu)$. Set*

$$E_n = \{x : A_j(f, x) > 0 \text{ for some } j \leq n\}.$$

Then

$$\int_{E_n} f \, \mathrm{d}\mu \geq 0$$

for all n.

Proof. Set

$$F_n(x) = \max\left(0, \sum_{i=0}^{j-1} f(T^i(x)) : j \leq n\right),$$

a non-decreasing sequence of functions. Observe

$$F_{n+1} = \max(0, f + F_n \circ T).$$

As

$$\max\left(f, f + f \circ T, \ldots, \sum_{i=0}^{n} f \circ T^i\right) \geq 0$$

on E_{n+1},

$$F_{n+1} = f + F_n \circ T$$

on E_{n+1} and

$$f = F_{n+1} - F_n \circ T$$

on E_{n+1}.

Thus

$$\int_{E_{n+1}} f\,d\mu = \int_{E_{n+1}} (F_{n+1} - F_n \circ T)\,d\mu.$$

Now $F_{n+1} = 0$ off E_{n+1} and $-F_n \circ T \leq 0$ everywhere so $F_{n+1} - F_n \circ T \leq 0$ off E_{n+1}.

So

$$\int_{E_{n+1}} f\,d\mu = \int_{E_{n+1}} (F_{n+1} - F_n T)\,d\mu \geq \int_X (F_{n+1} - F_n T)\,d\mu$$

$$= \int_X (F_{n+1} - F_n)\,d\mu \geq 0$$ ■

Corollary 3.3 *Setting* $E_\infty = \bigcup_{n=1}^{\infty} E_n$,

$$\int_{E_\infty} f\,d\mu \geq 0.$$ ■

Theorem 3.4 (Birkhoff ergodic theorem). *Let T be a measure-preserving transformation on a σ-finite measure space $(X, \mathscr{F}, \mu)$, and $f \in L^1(\mu)$. There exists an $\bar{f}$ with*

$$A_n(f, x) \to \bar{f}(x) \quad a.e.$$

Proof. For rational u and v let

$$E_{u,v} = \{x : \overline{\lim}\, A_n(f, x) > v > u > \underline{\lim}\, A_n(f, x) > 0\}. \tag{3.3}$$

If x belongs to no such $E_{u,v}$ then $\lim_{n\to\infty} A_n(f, x)$ exists (possibly $\pm\infty$). We will show $\mu(E_{u,v}) = 0$.

Assume $v > 0$, otherwise replace f by $-f$ and $-u > 0$. Assume $\mu(E_{u,v}) > 0$. From its definition, if $T(x) \in E_{u,v}$, then x is also. In other words, $T^{-1}(E_{u,v}) \subseteq E_{u,v}$, so we may, without loss of generality, assume $E_{u,v}$ is the entire measure space. Hence we can assume for all $x \in X = E_{u,v}$, for some n,

$$\frac{1}{n}\sum_{i-0}^{n-1} (f(T^i(x)) - v) > 0.$$

The function $f - v \notin L^1(\mu)$ if $\mu(X) = \infty$, but choosing a set $A \subset X$, $\mu(A) < \infty$, for each x

$$\frac{1}{n}\sum_{i=0}^{n-1} (f(T^i(x)) - v\chi_A(T^i(x))) > 0 \tag{3.4}$$

and $f - v\chi_A \in L^1(\mu)$. Hence by the maximal ergodic lemma

$$\int_X (f - v\chi_A)\,d\mu \geq 0$$

or

$$\int_X f \, d\mu \geq v\mu(A) \tag{3.5}$$

since X is the set E_∞ of Corollary 3.3. Letting A increase to all of X, $\|f\|_1 \geq v\mu(X)$ and as $v > 0$, $\mu(X) < \infty$. Now, using $\mu(X) < \infty$, $u - f \in L^1(\mu)$ and for some n,

$$u - \frac{1}{n}\sum_{i=0}^{n-1} f(T^i(x)) > 0.$$

By the maximal lemma

$$\int_X (u - f) \, d\mu \geq 0$$

or

$$\int_X f \, d\mu \leq u\mu(E_{u,v}). \tag{3.6}$$

This implies $\mu(E_{u,v}) = 0$ and we are done. ∎

Corollary 3.5 *Defining the map $L(f) = \bar{f}$, for $f \in L^1(\mu)$, $\|L(f)\|_1 \leq \|f\|_1$ and so $L(f)$ is a continuous projection from $L^1(\mu)$ onto the subspace of T-invariant L^1-functions.*

Proof As $\|A_n(g)\|_1 = \|g\|_1$ for $g \geq 0$, and since $A_n(f)$ converges pointwise to $\bar{f}$,

$$\int |L(f)| \, d\mu \leq \int L(|f|) \, d\mu \leq \underline{\lim} \int A_n(|f|) \, d\mu = \int |f| \, d\mu$$

and

$$\|L(f)\|_1 \leq \|f\|_1. \tag{3.7}$$

Notice that

$$\begin{aligned} L(f)(T(x)) &= \lim A_n(f, T(x)) \\ &= \lim \frac{1}{n}\sum_{i=1}^{n} f(T^i(x)) \\ &= \lim\left(\frac{n+1}{n} A_{n+1}(f, x) - \frac{f(x)}{n}\right) \\ &= L(f)(x). \end{aligned} \tag{3.8}$$

Thus $L(f)$ is T-invariant. As $\|L(f - g)\|_1 = \|L(f) - L(g)\|_1 \leq \|f - g\|_1$, L is a continuous projection onto the T-invariant L^1 functions. ∎

We now show that the only non-trivial convergence in the Birkhoff theorem is in the case of a finite invariant measure.

Corollary 3.6 *If $(X, \mathscr{F}, \mu)$ has no T-invariant subsets of finite measure, then $L(f) \equiv 0$ for all $f \in L^1(\mu)$.*

Proof If X has no T-invariant sets of finite measure, the only T-invariant L^1 function is identically equal to 0. ∎

Notice that X can be broken up into a subset X_∞ which has no T-invariant subsets of finite measure, and its complement which is at most a countable union of T-invariant sets of finite measure. On X_∞, Cesáro averages of L^1 functions converge to zero. The following corollary says that on a remaining piece the convergence is in L^1.

Corollary 3.7 *If $\mu(X) < \infty$, then $\|A_n(f) - L(f)\|_1 \to 0$.*

Proof Define

$$\mathscr{A} = \{f \in L^1(\mu) : \|A_n(f) - L(f)\|_1 \to 0\}.$$

As the operator L is a contraction in L^1, $\mathscr{A}$ is L^1-closed (if f_i is Cauchy so is $L(f_i)$).

If f is bounded then all $A_n(f)$ have the same bound and by the dominated convergence theorem, $A_n(f) \to L(f)$ in L^1. The only closed subspace of $L^1(\mu)$ that contains all bounded functions is $L^1(\mu)$ itself. ∎

Corollary 3.8 *If $(X, \mathscr{F}, \mu)$ is a probability space, T a measure-preserving transformation, and $f \in L^1(\mu)$, then $L(f) = E(f|\mathscr{I})$ where $\mathscr{I}$ is the algebra of T-invariant sets.*

Proof Using Exercise 2.12 all we need show is

(1) $L(f)$ is $\mathscr{I}$-measurable (which is true as L is T-invariant); and

(2) for any $A \in \mathscr{I}$

$$\int_A L(f)\,\mathrm{d}\mu = \int_A f\,\mathrm{d}\mu.$$

As A is T-invariant, $\int_A A_n(f)\,\mathrm{d}\mu = \int_A f\,\mathrm{d}\mu$ and as $A_n(f) \to L(f)$ in L^1, $\int_A L(f)\,\mathrm{d}\mu = \int_A f\,\mathrm{d}\mu$, and we are done. ∎

The power of the Birkhoff theorem thus lies where $\mu(X) < \infty$. We will now construct another development of the proof in the case where $(X, \mathscr{F}, \mu)$ is a Lebesgue probability space. The proof has a similar flavour to all pointwise convergence arguments, showing that sets $E_{u,v}$, where the value of the sequence oscillates infinitely often above v and below u, has measure 0. This time, though, instead of using a maximal lemma we use a Vitali type lemma to 'disjointify' segments of orbit on which the averages are above or below the appropriate bounds. Compare this argument with the martingale theorem of the previous chapter.

We state here the Vitali lemma we use and show how it proves the Birkhoff theorem. We postpone the proof of the lemma to the end of the chapter and on first reading, we recommend the reader omit its study. It is an extremely useful fact but the argument is quite technical.

We say $x \in X$ is a *periodic point* of *least period* n for T if $T^n(x) = x$, and n is the least such value. We say T is *non-periodic* if the collection of all periodic points has measure 0.

Theorem 3.9 (Backward Vitali lemma). *Suppose $(X, \mathscr{F}, \mu)$ is a Lebesgue probability space and T a non-periodic measure-preserving invertible map of X to itself.*

Suppose $A \subseteq X$ has $\mu(A) > 0$ and for every $x \in A$ we have sequences of measurable integer-valued functions $i_k(x) \leq 0 \leq j_k(x)$ with $\lim_{k \to \infty} (j_k(x) - i_k(x) + 1) = \infty$ for all $x \in A$. We can then for any $\varepsilon > 0$ find a subset $A' \subseteq A$ and measurable functions $i(x) \leq 0 \leq j(x)$, defined for $x \in A'$ where $(i(x), j(x)) = (i_k(x), j_k(x))$ for some k (depending on x) so that the sets $I(x) = \bigcup_{i=i(x)}^{j(x)} T^i(x)$ are pairwise disjoint and

$$\mu\left(A \Big\backslash \bigcup_{x \in A'} I(x)\right) < \varepsilon.$$

Notice the analogy between this result and the standard Vitali covering lemma. There are differences however. The most obvious one is that the intervals $I_k(x) = \bigcup_{i=i_k(x)}^{j_k(x)} T^i(x)$ are increasing on the orbits of T instead of decreasing on $\mathbb{R}$. The basic geometry of the proof, though, is the same.

Fig. 3.1 draws $\bigcup_{x \in A'} I(x)$ schematically. The set $I(x)$ cuts vertically through the figure. The intervals drawn correspond to disjoint measurable subsets of X, and cover all but ε in measure of X.

Theorem 3.10 (Rohlin Lemma). *Let T be non-periodic and as above. For any $n \in N$ and $\varepsilon > 0$ there is a set $F \subset X$ so that $F, T(F), \ldots, T^{n-1}(F)$ are disjoint and their union covers all but ε in measure of X.*

Proof For all $x \in X$, let $i_k(x) = 0, j_k(x) = nk - 1$. There is, by the backward Vitali lemma, a set $A' \subset X$ and measurable values $i(x)$, $j(x)$ defined on A', chosen from among the $i_k(x)$, $j_k(x)$ so that all $\bigcup_{i=i(x)}^{j(x)} T^i(x) = I(x)$ are disjoint and

$$\mu\left(\bigcup_{x \in A'} I(x)\right) > 1 - \varepsilon.$$

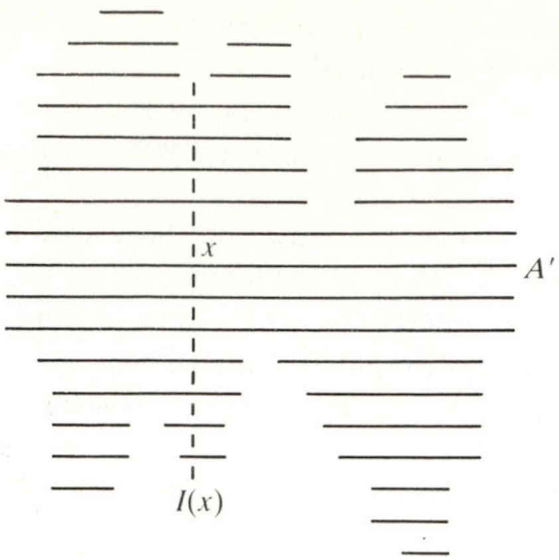

Fig. 3.1 Schematic of the backward Vitali lemma. The intervals represent disjoint measurable subsets, and the orbit sections $I(x)$ cut through the diagram vertically.

Now $i(x)$ and $j(x)$ must be of the form $i(x) = 0$ and $j(x) = nk(x) - 1$. Set

$$F = \bigcup_{x \in A'} \left(\bigcup_{k=0}^{k(x)-1} T^{nk}(x) \right). \tag{3.9}$$

Letting

$$I'(x) = \bigcup_{k=1}^{n-1} T^{i}(x) \quad \text{for } x \in F,$$

then

$$I(x) = \bigcup_{k=0}^{k(x)-1} I'(T^{nk}(x))$$

is a disjoint union. Thus for $x \in F$, the $I'(x)$ are pairwise disjoint and

$$\mu\left(\bigcup_{x \in F} I'(x) \right) = \mu\left(\bigcup_{x \in A'} I(x) \right) > 1 - \varepsilon. \tag{3.10}$$

Saying the $I(x)$ are pairwise disjoint is equivalent to saying F, $T(F)$, ..., $T^{n-1}(F)$ are pairwise disjoint. ∎

The Rohlin lemma is in fact much older than the backward Vitali lemma, and this is a very unusual and in fact difficult proof, in that it uses the much deeper backward Vitali lemma.

Exercise 3.1 Give a direct proof of the Rohlin lemma, not relying on the backward Vitali lemma.

Standard proofs usually rely heavily on the ordering of $\mathbb{Z}$, as did our first proof of the Birkhoff theorem, or look surprisingly like the proof of the backward Vitali lemma. In this sense the backward Vitali lemma can be viewed as a generalized Rohlin lemma. In fact for non-periodic actions of general amenable groups the backward Vitali lemma remains true with only slight changes (Følner sets replace intervals and complete disjointness is not obtained) whereas the Rohlin lemma as stated requires the existence of tiling sets (Ornstein and Weiss 1980).

We now use the backward Vitali lemma to re-prove the Birkhoff theorem. Suppose we have a set F and measurable functions $i(x) \leq 0 \leq j(x)$ so that the sets $I(x) = \bigcup_{i=i(x)}^{j(x)} T^i(x)$ are disjoint as x varies over F. A basic fact to keep in mind is that for any $f \in L^1(\mu)$ we get

$$\int_{\bigcup_{x \in F} I(x)} f \, d\mu = \int_F \sum_{i=i(x)}^{j(x)} f(T^i(x)) \, d\mu. \tag{3.11}$$

Theorem 3.11 *Let T be a measure-preserving, invertible transformation on $(X, \mathcal{F}, \mu)$, a Lebesgue probability space. Let $f \in L^1(\mu)$ and as before set*

$$A_n(f, x) = \frac{1}{n} \sum_{i=0}^{n-1} f(T^i(x)).$$

For almost all $x \in X$, $A_n(f, x)$ converges.

We can, as usual, identify the limit function. Let

$$\mathcal{I} = \{A \in \mathcal{F} : \mu(A \Delta T^{-1}(A)) = 0\}.$$

This is a σ-algebra, the σ-algebra of T-invariant (mod 0) *sets. For almost all $x \in X$,*

$$A_n(f, x) \xrightarrow{n} E(f \mid \mathcal{I}) = \bar{f}(x). \tag{3.12}$$

Proof We first handle periodic points. Let $X = \bigcup_{i=0}^{\infty} X_n \cup X_\infty$ where $x \in X_n$ iff x is of least period n. All the X_n are in $\mathcal{I}$, and $\mathcal{I}$ restricted to X_n separates the n-point orbits. Thus both $E(f \mid \mathcal{I})$ and $\lim_{k \to \infty} f_k$ are equal to $(1/n) \sum_{i=0}^{n-1} f(T^i(x))$ on X_n.

We are left with X_∞, on which T is non-periodic. Renormalizing the measure, we have reduced the problem to T a non-periodic transformation of $(X, \mathcal{F}, \mu)$.

Define

$$\hat{f}(x) = \lim_{n \to \infty} \sup(A_n(f, x)),$$

taking values in $R \cup \{\infty, -\infty\}$. As $\hat{f}(T(x)) = \hat{f}(x)$, $\hat{f}$ is $\mathcal{I}$ measurable.

Let $E \in \mathcal{I}$. We wish to show that $\int_E \hat{f} \, d\mu = \int_E f \, d\mu$, and hence $\hat{f} = E(f \mid \mathcal{I})$ a.e. As the sets where $\hat{f} > 0$ and $\hat{f} \leq 0$ are in $\mathcal{I}$ we can, without loss of generality, assume $\hat{f} \geq 0$.

Let

$$E_\infty = \{x \in E : \hat{f}(x) = \infty\},$$
$$E_M = \{x \in E : \hat{f}(x) \le M\}. \tag{3.13}$$

Fix $\varepsilon > 0$ and $M > 0$ and we can measurably select functions $i_k(x) = 0$, $j_k(x) \xrightarrow{k} \infty$ so that

(1) $A_{j_k(x)+1}(f, x) > \dfrac{1}{\varepsilon}$ if $x \in E_\infty$;

(2) $|A_{j_k(x)+1}(f, x) - \hat{f}(x)| < \varepsilon$ for $x \in E_M$. (3.14)

Use the backward Vitali lemma to find a set $F \subseteq X$ and functions $i(x) = 0$, $j(x) = j_k(x)$ for some k so that all the $I(x) = \bigcup_{i=i(x)}^{j(x)} T^i(x)$ are disjoint and $B = \bigcup_{x \in F} I(x)$ has $\mu(B) > 1 - \varepsilon$. As E_∞, $E_M \in \mathscr{I}$, if $x \in E_\alpha$, then $I(x) \subseteq E_\alpha$. Thus for E_∞,

$$\int_{E_\infty \cap B} f \, d\mu = \int_{E_\infty \cap F} (j(x) + 1) A_{j(x)+1}(f, x) \, d\mu$$
$$\ge \frac{1}{\varepsilon} \int_{E_\infty \cap F} (j(x) + 1) \, d\mu = \frac{\mu(E_\infty \cap B)}{\varepsilon}.$$

So

$$\varepsilon \| f \|_1 \ge \mu(E_\infty \cap B).$$

Letting $\varepsilon \to 0$, $\mu(E_\infty) = 0$. Thus $E = \lim_M E_M$ a.s. so $0 \le \hat{f} < \infty$ a.e.

Now by (3.11)

$$\varepsilon \ge \varepsilon\mu(E_M \cap B) \ge \varepsilon \int_{E_M \cap F} (j(x) + 1) \, d\mu$$
$$\ge \int_{E_M \cap F} |\hat{f}(x) - A_{j(x)+1}(f, x)| (j(x) + 1) \, d\mu$$
$$\ge \left| \int_{E_M \cap F} (\hat{f}(x) - A_{j(x)+1}(f, x))(j(x) + 1) \, d\mu \right|$$
$$\ge \left| \int_{E_M \cap F} \hat{f} - f \, d\mu \right|. \tag{3.15}$$

Letting $\varepsilon \to 0$, $\int_{E_M} (\hat{f} - f) \, d\mu = 0$ as $\hat{f}$ is bounded on E_M. Letting $M \to \infty$, by monotone convergence $\int_E \hat{f} \, d\mu = \int_E f \, d\mu$. Using $-f$ on the negative part of $\hat{f}$ we get the result for all $E \in \mathscr{I}$.

Replacing f by $-f$, $\liminf_{n \to \infty} f_n(x) = E(f | \mathscr{I})$ a.s. hence the limit exists a.e. and is $E(f | \mathscr{I})$. ■

3.3 Proof of the backward Vitali lemma

We now digress to prove the backward Vitali lemma. On first reading we recommend skipping this proof, jumping forward to Corollary 3.16.

Lemma 3.12 *If the set of all periodic points for T of least period less than N has measure 0, and $\mu(A) > 0$, then there is a subset $A' \subseteq A$, $\mu(A') > 0$ and all the sets A', $T(A'), \ldots, T^{N-1}(A')$ are disjoint.*

Proof We prove the contrapositive. Suppose $\mu(A) > 0$ and for all $A' \subseteq A$, for some $0 < n < N$,

$$T^n(A') \cap A' \neq \varnothing.$$

As we could always delete sets of measure 0 from A', we must have an n with

$$\mu(T^n(A') \cap A') > 0.$$

Let $n(A')$ be the least $n > 0$ with $\mu(T^n(A') \cap A') > 0$. If $A'' \subseteq A'$ then $n(A'') \geq n(A')$.

Let

$$N_0 = \max_{\substack{A' \subseteq A \\ \mu(A') > 0}} (n(A')) < N.$$

There is a subset $A' \subseteq A$, $\mu(A') > 0$ and $n(A') = N_0$, so A', $T(A'), \ldots, T^{N_0-1}(A')$ are all disjoint.

Let $A'' = A' \backslash T^{N_0}(A')$. As A'', $T(A''), \ldots, T^{N_0-1}(A'')$, $T^{N_0}(A'')$ are all disjoint, $n(A'') > N_0$ so $\mu(A'') = 0$, i.e.,

$$A' = T^{N_0}(A') \quad \text{a.s.}$$

This must also be true of any subset of A'.

Let P_i be a refining tree of partitions of A'. For any subset $S \in P_i$,

$$T^{N_0}(S) = S \quad \text{a.s.}$$

Delete from A' a T^{N_0}-invariant subset of measure 0 so that on what remains (A''),

$$T^{N_0}(S \cap A'') = S \cap A''$$

for all $S \in P_i$. For any chain of sets $\{c_i\}$ in the tree P_i, that descends to a point $x \in A''$, it follows that

$$T^{N_0}(x) = x.$$

Thus every point of A'' is a periodic point of period $N_0 < N$. ∎

Corollary 3.13 *If T is non-periodic, then for any subset A, $\mu(A) > 0$ and $i \leq 0 \leq j$, there is a subset $A' \subseteq A$, $\mu(A') > 0$ with $T^i(A')$, $T^{i+1}(A'), \ldots, T^j(A')$ all disjoint.* ∎

Exercise 3.2 Let $P_n = \{x : T^n(x) = x \text{ and } n \text{ is the least such}\}$. If $\mu(P_n) > 0$ show that $P_n = \bigcup_{i=0}^{n-1} A_i$ where the A_i are disjoint, $T^{-1}(A_i) = A_{(i-1) \bmod n}$. This shows what the periodic part of T looks like.

Exercise 3.3. Suppose T is ergodic but T^n is *not*, for some $n > 1$. Show that there is a value k dividing n, $k \neq 1$, and disjoint sets $A_0, A_1, \ldots, A_{k-1}$ with $x = \bigcup_{i=0}^{k-1} A_i$ a.s., $T^{-1}(A_i) = A_{(i-1) \bmod k}$ and T^k acting on A_0 is ergodic.

We are now ready for our two most technical steps. Notice the parallel between them and the proof of the standard Vitali lemma of Chapter 2. The use of one-dimensional geometry is the same. Because our sequences of intervals nest outward instead of inward, we must use much more explicit constructions than in the classical Vitali lemma.

Lemma 3.14 *Let $(X, \mathcal{F}, \mu)$ be a Lebesgue probability space, T a measure-preserving invertible non-periodic map of X to itself. On a set $A \subset X$, $\mu(A) > 0$, we are given bounded integer-valued measurable functions $i(x) \leq 0 \leq j(x)$. There is then a measurable subset $A' \subset A$, so that for all $x \in A'$, the sets $I(x) = \bigcup_{i=i(x)}^{j(x)} T^i(x)$ are disjoint (i.e., $I(x) \cap I(x') = \varnothing$ unless $x = x'$) and $\mu(\bigcup_{x \in A} I(x)) \geq \mu(A)/3$.*

Proof Let $A_{i,j} = \{x \in A : i(x) = i, j(x) = j\}$, a measurable set. As $i(x), j(x)$ are bounded there are only finitely many such sets. Order the sets $A_{i,j}$ as $A_{i(1),j(1)} \ldots A_{i(M),j(M)}$ so that

$$j(k+1) - i(k+1) \leq j(k) - i(k), \tag{3.16}$$

i.e., the lengths of the blocks $I(x)$, $x \in A_{i(k),j(k)}$ are non-increasing in k. Let $n(k) = j(k) - i(k) + 1$ be this length.

We construct a sequence of sets $A'_k \subseteq A_{i(k),j(k)}$ inductively. A' will be the union of these sets.

Let

$\mathcal{A}_1 = \{A' \subseteq A_{i(1),j(1)} : \mu(A') > 0, \text{ and } T^{i(1)}(A'), T^{i(1)+1}(A'), \ldots, T^{j(1)}(A') \text{ are disjoint}\}$.

By the previous lemma $\mathcal{A}_1 \neq \phi$.

If $A'_1 \subseteq A'_2 \subseteq \cdots \subseteq A'_k \subseteq \cdots$ are all in $\mathcal{A}_1$ then so is $\bigcup_{i=1}^{\infty} A'_i$. Thus there must be an a.s.-maximal set under containment $A'_1 \in \mathcal{A}_1$, i.e., if $A'_1 \subseteq A'' \in \mathcal{A}_1$ then $\mu(A'') = \mu(A'_1)$. (This is *not* done with the axiom of choice. Rather the sequence is constructed to approach the maximal available measure.)

We claim

$$A_{i(1),j(1)} \subseteq \bigcup_{i=i(1)-n(1)}^{j(1)+n(1)} T^i(A'_1) \quad \text{a.s.} \tag{3.17}$$

If not, then

$$B = A_{i(1),j(1)} \Big\backslash \bigcup_{i=i(1)-n(1)}^{j(1)+n(1)} T^i(A'_1)$$

would be of positive measure and hence contain a subset B', $\mu(B') > 0$ and $T^{i(1)}(B'), T^{i(1)+1}(B'), \ldots, T^{j(1)}(B')$ all disjoint and disjoint from $\bigcup_{i=i(1)}^{j(1)} T^i(A'_1)$. But then $B' \cup A'_1 \in \mathscr{A}_n$, conflicting with maximality of A'_1.

Arguing now by induction, to complete the proof suppose we have subsets A'_u, $1 \le u < k$, with all sets of the form

$$T^i(A'_u), \quad i(u) \le i \le j(u)$$

pairwise disjoint and

$$\bigcup_{u=1}^{k-1} A_{i(u),j(u)} \subseteq \bigcup_{u=1}^{k-1} \left(\bigcup_{i=i(u)-n(u)}^{j(u)+n(u)} T^i(A'_u) \right) \quad \text{a.s.} \tag{3.18}$$

Set

$$\bar{A}_k = A_{i(k),j(k)} \bigcup_{u=1}^{k-1} \left(\bigcup_{i=i(u)-n(u)}^{j(u)+n(u)} T^i(A'_u) \right).$$

If $\mu(\bar{A}_k) = 0$, setting $A'_k = \varnothing$, the induction extends to k.

Otherwise $\mu(\bar{A}_k) \ne 0$. Let

$\mathscr{A}_k = \{A' \subseteq \bar{A}_k : \mu(A') > 0$, and $T^{i(k)}(A'), T^{i(k)+1}(A'), \ldots, T^{j(k)}(A')$ are disjoint$\}$.

As before $\mathscr{A}_k$ is non-empty and must contain an a.s.-maximal element under containment. Call it A'_k, and

$$\bar{A}_k \subseteq \bigcup_{i=i(k)-n(k)}^{j(k)+n(k)} T^i(A'_k). \tag{3.19}$$

As $A'_k \subset \bar{A}_k$, and $n(k) \le n(u)$ for $u < k$, $\bigcup_{i=i(k)}^{j(k)} T^i(A'_k)$ is disjoint from $\bigcup_{u=1}^{k-1} (\bigcup_{i=i(u)}^{j(u)} T^i(A'_u))$, so all sets $T^i(A'_u)$, $1 \le u \le k$, $i(u) \le i \le j(u)$, are pairwise disjoint. Combining (3.18) and (3.19),

$$\bigcup_{u=1}^{k} A_{i(u),j(u)} \subseteq \bigcup_{u=1}^{k-1} \left(\bigcup_{i=i(u)-n(u)}^{j(u)+n(u)} T^i(A'_u) \right). \tag{3.20}$$

Continue the induction through M and

$$A \subseteq \bigcup_{u=1}^{M} \left(\bigcup_{i=i(u)-n(u)}^{j(u)+n(u)} T^i(A'_u) \right).$$

Set $A' = \bigcup_{u=1}^{M} A'_u$ and

$$\bigcup_{x \in A'} I(x) = \bigcup_{u=1}^{M} \left(\bigcup_{i=i(u)}^{j(u)} T^i(A'_u) \right), \tag{3.21}$$

a disjoint union. To see that disjointness of this union is equivalent to disjointness of the sets $I(x)$, suppose $x, x' \in A'$ and $I(x) \cap I(x') \neq \emptyset$. Then $x \in A_k$, $x' \in A_{k'}$ and for some $i(k) \leq i \leq j(k)$, $i(k') \leq i' \leq j(k')$, $T^i(x) = T^{i'}(x')$ so $T^i(A_k) \cap T^{i'}(A_{k'}) \neq \emptyset$ and $k = k'$, $i = i'$ and $x = x'$. As

$$A \subseteq \bigcup_{u=1}^{M} \left(\bigcup_{i=i(u)-n(u)}^{j(u)+n(u)} T^i(A'_u) \right),$$

$$\mu(A) \leq \sum_{u=1}^{M} 3 \sum_{i=i(u)}^{j(u)} \mu(T^i(A'_u)) = 3\mu\left(\bigcup_{x \in A'} I(x) \right). \qquad \blacksquare$$

The conclusion of the next corollary is precisely that of the backward Vitali lemma. The hypotheses are stronger though. Conditions (1) and (2) are growth conditions on the orbit intervals. We shall see later that they are obtained by dropping to a subsequence.

Corollary 3.15 *For $(X, \mathscr{F}, \mu)$ and T as in the previous lemma, and $1 \geq \varepsilon > 0$, suppose we have a set $A \subset X$, $\mu(A) > 0$ and for $x \in A$, $k = 1, \ldots, M$ we have bounded integer-valued functions $i_k(x) \leq 0 \leq j_k(x)$ so that*

(1) $\dfrac{\varepsilon M}{12} > 1$, *and*

(2) $$\sup_{x \in A} (j_{k+1}(x) - i_{k+1}(x) + 1) \leq \frac{\varepsilon}{4} \inf_{x \in A} (j_k(x) - i_k(x) + 1). \tag{3.22}$$

It follows that there is a subset $A' \subset A$ and measurable functions $i(x) \leq 0 \leq j(x)$ for $x \in A'$ with $(i(x), j(x)) = (i_k(x), j_k(x))$ for some k depending on x. The sets

$$I(x) = \bigcup_{i=i(x)}^{j(x)} T^i(x)$$

are pairwise disjoint for $x \in A'$ and

$$\mu\left(A \setminus \bigcup_{x \in A'} I(x) \right) < \varepsilon.$$

Proof Let

$$i'_k(x) = i_k(x) - \sup_{x \in A} (j_{k+1}(x) - i_{k+1}(x) + 1)$$

$$j'_k(x) = j_k(x) + \sup_{x \in A} (j_{k+1}(x) - i_{k+1}(x) + 1). \tag{3.23}$$

Use the previous lemma inductively to get a subset $A'_1 \subset A$ with $I'(x) = \bigcup_{i=i'_1(x)}^{j'_1(x)} T^i(x)$ pairwise disjoint for $x \in A'_1$ and

$$\mu\left(\bigcup_{x \in A'} I'_1(x) \right) \geq \frac{1}{3} \mu(A).$$

Set $A_2 = A \backslash \bigcup_{x \in A_1'} I'(x)$. If $\mu(A_2) > 0$ repeat the procedure using $i_2(x), j_2(x)$ to find $A_2' \subseteq A_2$, set $I'(x) = \bigcup_{i=i_2'(x)}^{j_2'(x)} T^i(x)$ pairwise disjoint for $x \in A_2'$ and

$$\mu\left(\bigcup_{x \in A_2'} I'(x)\right) \geq \frac{1}{3}\mu(A_2).$$

Continuing inductively, set

$$A_k = A \Big\backslash \bigcup_{i=1}^{k-1} \bigcup_{x \in A_i'} I'(x) = A_{k-1} \Big\backslash \bigcup_{x \in A_{k-1}'} I'(x). \tag{3.24}$$

If $\mu(A_k) > 0$ repeat the procedure using $i_k(x), j_k(x)$ to find $A_k' \subseteq A_k$. Set $I'(x) = \bigcup_{i=i_k'(x)}^{j_k'(x)} T^i(x)$, pairwise disjoint for $x \in A_k'$ and

$$\mu\left(\bigcup_{x \in A_k'} I'(x)\right) \geq \frac{1}{3}\mu(A_k). \tag{3.25}$$

Continue through the M steps. Set $A' = \bigcup_{k=1}^{M} A_k'$, a disjoint union. For $x \in A'$, x is in a unique A_k' so set $i(x) = i_k(x), j(x) = j_k(x)$, and now

$$I(x) = \bigcup_{i=i(x)}^{j(x)} T^i(x).$$

If $x, x' \in A', x \neq x'$ and $I(x) \cap I(x') \neq \varnothing$, then as for all y, $I(y) \subset I'(y)$, we must have $x \in A_{k_1}, x' \in A_{k_2}, k_1 \neq k_2$. Then $x = T^n(x')$ where

$$|n| \leq (j_{k_1}(x) - i_{k_1}(x) + 1) + (j_{k_2}(x') - i_{k_2}(x') + 1)$$

and either $x \in I'(x')$ or $x' \in I'(x)$, which is a conflict. Hence

$$I(x) \cap I(x') = \varnothing.$$

From this disjointness,

$$\mu\left(\bigcup_{x \in A'} I(x)\right) = \sum_{k=1}^{M} \mu\left(\bigcup_{x \in A_k'} I(x)\right)$$

and

$$\begin{aligned}
\mu\left(\bigcup_{x \in A_k'} I(x)\right) &= \int_{A_k'} (j_k(x) - i_k(x) + 1)\,\mathrm{d}\mu \\
&\geq \left(1 - \frac{\varepsilon}{2}\right) \int_{A_k'} (j_k'(x) - i_k'(x) + 1)\,\mathrm{d}\mu \\
&\geq \left(1 - \frac{\varepsilon}{2}\right) \mu\left(\bigcup_{x \in A_k'} I'(x)\right) \geq \left(1 - \frac{\varepsilon}{2}\right) \frac{\mu(A_k)}{3}.
\end{aligned} \tag{3.26}$$

Letting $A_{M+1} = A \backslash \bigcup_{k-1}^{M} \bigcup_{x \in A_k'} I(x)$, we have $A \supseteq A_2 \supseteq A_3 \supseteq \cdots \supseteq A_{M+1}$. Suppose $\mu(A_{M+1}) > \varepsilon/2$, i.e., $\mu(A_k) > \varepsilon/2$ for all k. Then

$$\mu\left(\bigcup_{x \in A'} I(x)\right) \geq M\left(1 - \frac{\varepsilon}{2}\right)\frac{\varepsilon}{6} > \frac{M\varepsilon}{12} > 1,$$

a conflict. Thus

$$\mu(A_{M+1}) \le \frac{\varepsilon}{2}$$

and

$$\begin{aligned}\mu\left(A\Big\backslash \bigcup_{x\in A'} I(x)\right) &\le \mu(A_{M+1}) + \mu\left(\bigcup_{x\in A'} (I'(x)\backslash I(x))\right)\\ &\le \frac{\varepsilon}{2} + \frac{\varepsilon}{2}\int_{A'} (j(x) - i(x) + 1)\,\mathrm{d}\mu\\ &= \frac{\varepsilon}{2} + \frac{\varepsilon}{2}\left(\bigcup_{x\in A'} I(x)\right) \le \varepsilon. \qquad (3.27)\ \blacksquare\end{aligned}$$

To finish the proof of the backward Vitali lemma, what we need to do is to see how to obtain boundedness of the functions i_k and j_k and conditions (3.22) of Corollary 3.15.

Proof of Theorem 3.9 (Backward Vitali Lemma) Having fixed $\varepsilon > 0$, select M so that $M\varepsilon/24 > 1$. We define new sequences of functions $\hat{i}_k$ and $\hat{j}_k$ inductively on successively reduced domains in A. Here are steps 1 and 2. Let $\hat{i}_1(x) = i_1(x), \hat{j}_1(x) = j_1(x)$. Delete at most $\varepsilon/4$ in measure of A to obtain a subset A_1 with $\hat{i}_1$ and $\hat{j}_1$ bounded on A_1. Select $k(x)$ measurably so that for all $x \in A$,

$$\sup_{x\in A_1} (\hat{j}_1(x) - \hat{i}_1(x) + 1) \le \frac{\varepsilon}{8}(j_{k(x)}(x) - i_{k(x)}(x) + 1).$$

Let $\hat{j}_2(x) = j_{k(x)}(x), \hat{i}_2(x) = i_{k(x)}(x)$. Delete from A_1 at most $\varepsilon/8$ in measure of A_1 to obtain a subset A_2 with $\hat{i}_2, \hat{j}_2$ bounded on A_2. Suppose inductively we have found subsets $A \supseteq A_1 \supseteq A_2 \supseteq \cdots \supseteq A_{k-1}$, $\mu(A_{u+1}) \ge \mu(A_u) - \varepsilon/2^{j+1}$, and $(\hat{i}_u(x), \hat{j}(x))$, $u = 1, \ldots, k-1$ are from among the $(i_v(x), j_v(x))$ (v depending measurably on x), are boundary and satisfy

$$\sup_{x\in A_{k-1}} (\hat{j}_u(x) - \hat{j}_u(x) + 1) \le \frac{\varepsilon}{4} \inf_{x\in A_{k-1}} (\hat{j}_{u+1}(x) - \hat{i}_{u+1}(x) + 1) \quad 1 \le u < k. \tag{3.28}$$

Select $k(x)$ measurably so that for all $x \in A_{k-1}$,

$$\frac{\varepsilon}{8}(j_{k(x)}(x) - i_{k(x)}(x)) \ge \sup_{x\in A_{k-1}} (\hat{j}_{k-1}(x) - \hat{i}_{k-1}(x) + 1). \tag{3.29}$$

Set $\hat{j}_k(x) = j_{k(x)}(x)$, $\hat{i}_k(x) = i_{k(x)}(x)$ and delete at most $\varepsilon/2^{k+1}$ in measure of A_{k-1} to obtain A_k on which both $\hat{j}_k(x)$ and $\hat{i}_k(x)$ are bounded.

Continue through M steps. Notice

$$\mu(A_M) \ge \mu(A) - \sum_{i=1}^{M} \frac{\varepsilon}{2^{j+1}} > \mu(A) - \frac{\varepsilon}{2}.$$

The previous Corollary 3.15 applies to the sequence of functions $\hat{i}_k(x), \hat{j}_k(x)$ in reverse order and the set A_M, with error $\varepsilon/2$. The conclusion of the corollary gives us the set $A' \subseteq A_M \leqslant A$ and $(i(x), j(x))$ from among $(\hat{i}_u(x), \hat{j}_u(x))$ which are from among the $(i_k(x), j_k(x))$ with $I(x) = \bigcup_{i=i(x)}^{j(x)} T^i(x)$ pairwise disjoint and

$$\mu\left(A \Big\backslash \bigcup_{x \in A'} I(x)\right) \leq \mu\left(A_M \Big\backslash \bigcup_{x \in A'} I(x)\right) + \frac{\varepsilon}{2} \leq \varepsilon. \qquad \blacksquare$$

Exercise 3.4 Formulate and prove a version of Theorem 3.9 in $\mathbb{Z}^n$. Can you do this for more general intervals than cubes?

3.4 Consequences of the Birkhoff theorem

As we know from Chapter 1, a map T is called *ergodic* if $\mathscr{I}$, the σ-algebra of T-invariant sets (mod 0) consists only of sets of measure 0 and 1. In this case the Cesáro averages f_n must converge a.e. to $\int f\,\mathrm{d}\mu$. We want a test for ergodicity that requires we check it on only countably many sets. The Birkhoff theorem provides the tool.

Corollary 3.16 *If $\{A_i\}$ is a countable collection of sets L^1 dense in the collection of all sets and*

$$\frac{1}{n}\sum_{i=1}^{n-1} \chi_{A_j}(T^i(x)) \to \mu(A_j) \tag{3.30}$$

for all j for a.e. x then T is ergodic.

Proof Let A be a measurable invariant set and A_j any from our collection. Now as A is invariant

$$\begin{aligned}\mu(A)\mu(A_j^c) &= \int_A \mu(A_j^c)\,\mathrm{d}\mu \\ &= \lim_{n\to\infty}\left(\frac{1}{n}\sum_{i=0}^{n-1}\int_A \chi_{A_j^c}(T^i(x))\,\mathrm{d}\mu\right) \\ &= \lim_{n\to\infty}\frac{1}{n}\sum_{i=0}^{n}\int \chi_{T^{-i}(A\cap A_j^c)}(x)\,\mathrm{d}\mu \\ &= \mu(A \cap A_j^c).\end{aligned} \tag{3.31}$$

Selecting A_j so $\mu(A\,\Delta\,A_j) \to 0$,

$$\mu(A)\mu(A^c) = 0$$

and $\mu(A) = 0$ or 1. $\blacksquare$

Often it is the case that T is given as a homeomorphism of a compact metric space without an invariant measure. The Birkhoff theorem can be used to obtain invariant measures, or at least information about them. The following exercises examine some parts of this issue.

Exercise 3.5 Throughout, suppose T is a homeomorphism of a compact metric space X. As usual, $A_n(f, x) = [f(x) + \cdots + f(T^{n-1}x)]/n$ for $f \in C_{\mathbb{R}}(X)$.

1. *Unique Ergodicity*. Suppose, for each f, there is a constant $L(f)$ so that $A_n(f, x) \to L(f)$ for all x in X. Then there is a unique invariant probability measure μ on X (with respect to the transformation T, and the Borel sets in X), namely the μ for which $L(f) = \int f \, d\mu$. Moreover, $A_n(f, x)$ converges to $L(f)$ uniformly in x.

The proof follows from the exercises below.

(a) $L(f)$ is a positive linear functional on $C_{\mathbb{R}}(X)$, and $L(1) = 1$. Moreover, $L(f \circ T) = L(f)$. By the Riesz representation theorem, there is a μ with $\mu(X) = 1$, $\mu \geq 0$, $L(f) = \int f \, d\mu$, and $\mu(T^{-1}E) = \mu(E)$ for all Borel E.

(b) If ν is an invariant probability measure then $\int A_n(f) \, d\nu = \int f \, d\nu$ and $\int A_n(f) \, d\nu \to \int L(f) \, d\nu = L(f) = \int f \, d\mu$. (You don't need any ergodic theorem here, Lebesgue dominated convergence will do.) Thus $\mu = \nu$ (cite a reason).

(c) If ν is a signed invariant measure (or, equivalently, a continuous linear functional on $C_{\mathbb{R}}(X)$) then $\nu = \tau\mu$ for some $\tau \in \mathbb{R}$. (Look up the appropriate decomposition theorem, including uniqueness.)

(d) Let $V = \{f : f = g - g \circ T + k, g \in C_{\mathbb{R}}, k \in \mathbb{R}\}$. Then V is dense in $C_{\mathbb{R}}(X)$ (in the norm $\|\cdot\|_\infty$). (Use (c), and an appropriate version of the Hahn–Banach theorem.)

(e) Use (d) to deduce $A_n(f) \to L(f)$ uniformly for all $f \in C_{\mathbb{R}}(X)$.

2. *Example*: Let $X = [0, 1]$ with the identification $0 = 1$. Let $Tx = x^2$ (so $T^n x = x^{(2^n)}$). Observe $T^n x \to 0$, for all $x \in X$, but *not* uniformly. Observe $A_n(f) \to f(0)$ uniformly for all continuous f.

3. *Example*: As an example of a map not uniquely ergodic, let T be an ergodic automorphism (or non-ergodic for that matter) of the d-dimensional torus $\mathbb{R}^d/\mathbb{Z}^d$, i.e., multiplication by an integer matrix of determinant ± 1. Show that the periodic points are dense (identify them!) and conclude there are many invariant measures.

Exercise 3.6 For each invariant probability measure μ on X, we have a linear functonal $L \in C^*_{\mathbb{R}}(X)$ satisfying $L \geq 0$ (i.e., $f \geq 0 \to L(f) \geq 0$), $L(1) = 1$, and $L(f \circ T) = L(f)$, and, conversely, such an L creats a μ. Note $\|L\| = 1 < \infty$ follows from the above.

1. Show the collection K of such L is a weak* closed subset of the unit ball in $C_{\mathbb{R}}(X)$, and thus weak* compact. Show the set is also convex. Conclude there are extreme points.

2. Show that T is ergodic with respect to the invariant measure μ if and only if L (corresponding to μ) is extreme in K.

Notes: If you can't make anything of the terminology in these last exercises, forget it for now. However, Sections 6 and 7 in Chapter 10 of Royden (1968), are relevant in 1. For 2, $\leftarrow$ follows easily once you form the contrapositive. The other direction needs the Radon–Nikodym theorem.

3.5 Disintegrating a measure space over a factor algebra

We now take leave of ergodic theorems and return to the constructive methods of Chapter 1. What we will show is that any measure-preserving map can be *disintegrated* into ergodic components. The idea here is to construct a map $\varphi : X \to [0,1] \times [0,1]$ defined a.e. and measurable onto a.a. of the unit square, taking μ to Lebesgue measure, $\mathscr{F}$ to the Lebesgue sets and $\mathscr{I}$ to the algebra of vertical sets $A \times [0,1]$. The measure spaces $\{x\} \times [0,1]$ with one-dimensional Lebesgue measure provide the desired disintegration and Fubini's theorem, the basic integration tool for re-integrating the decomposition. These ideas originate in the work of Rohlin (1966) and Stone (1950).

We will in fact consider a case slightly more general than the invariant algebra $\mathscr{I}$. $\mathscr{A}$ can be any T-invariant complete sub-σ-algebra of $\mathscr{F}$ containing $\mathscr{I}$. Here are two examples of such a situation.

Example 1 X is $[0,1] \times \{1,2,\ldots,n\}$, μ is the direct product of Lebesgue measure and normalized counting measure, and $\mathscr{F}$ is the Lebesgue algebra. T is a bimeasurable measure-preserving map of X that maps a fibre $\{x\} \times \{1,2,\ldots,n\}$ to another such fibre, measurably and preserving μ. $\mathscr{A}$ is the completion of the algebra of vertical sets $A \times \{1,2,\ldots,n\}$ where A is Lebesgue measurable in $[0,1]$.

Example 2 X is $[0,1] \times [0,1]$, μ is Lebesgue measure, $\mathscr{F}$ is the Lebesgue algebra, T is a bimeasurable measure-preserving map of X, taking fibres $\{x\} \times [0,1]$ to fibres bimeasurably preserving Lebesgue measure on the fibres. $\mathscr{A}$ is the completion of the algebra of vertical sets $A \times [0,1]$.

Exercise 3.7 Construct explicit examples of both of these types.

In both examples μ can be written as an integral over $[0,1]$ of fibre measures, (either counting measure/n or Lebesgue measure) by Fubini's theorem. Letting $f(x)$ be the fibre over x, μ_x the fibre measure and $\mathscr{F}_x$ the measurable subsets of the fibre, our examples have:

(1) $T:(f(x),\mathscr{F}_x,\mu_x)\to(f(T(x)),\mathscr{F}_{T(x)},\mu_{T(x)})$ bimeasurably and preserving measure;

(2) for $A\in\mathscr{F}$, for a.e. x, $\mu_x(A)=E(A|\mathscr{A})$; and

(3)
$$\mu(A)=\int_0^1 \mu_x(A)\,d\ell. \tag{3.32}$$

A more general example would be a weighted average of examples like 1 and 2. Of course (3.32) would still hold.

If we have a T-invariant algebra $\mathscr{A}$, and can find a bimeasurable measure-preserving map from $(X,\mathscr{F},\mu)$ to almost all of a weighted average of examples of type 1 and 2, then we can pull back the fibres, fibre algebras and fibre measures to X. We call such a collection $(f(x),\mathscr{F}_x,\mu_x)$ a *disintegration* of $(X,\mathscr{F},\mu)$ over the algebra $\mathscr{A}$.

Theorem 3.17 *If $\mathscr{A}$ is a non-atomic T-invariant complete sub-σ-algebra of $\mathscr{F}$, $\mathscr{I}\subset\mathscr{A}$ (T, of course, bimeasurable and measure-preserving) then $(X,\mathscr{F},\mu)$ can be disintegrated over $\mathscr{A}$.*

Note: We can leave out the non-atomic condition by putting atoms in the first coordinate. The $\mathscr{I}\subset A$ condition can also be removed but then requires more elaborate possible cases. The former issue we leave to the reader, the latter never arises in our work. What we are proving is a special case of the Rohlin–Stone decomposition of a measure space over a factor algebra (Rohlin 1966, Stone 1950).

Proof Our problem is to construct the requisite map to a weighted average of examples of types 1 and 2.

Let $\{P_i\}$ be a generating tree of partitions for $\mathscr{F}$, with $T(P_i)$ and $T^{-1}(P_i)$ both P_{i+1} measurable.

Let $\{Q_i\}$ be a generating tree of partitions for $\mathscr{A}$ with the same property. Remember, $\mathscr{A}$ is a Lebesgue algebra (Theorem 2.2).

For a partition $P=\{S_1,\ldots,S_k\}$ and algebra B we write

$$D(P|B)=(E(\chi_{s_1}|B),E(\chi_{s_2}|B),\ldots,E(\chi_{s_k}|B)), \tag{3.33}$$

a probability vector valued, B measurable function defined a.e.

We know, for a.e. x, for all i,

$$D(P_i|Q_j)(x)\xrightarrow{j}D(P_i|\mathscr{A})(x).$$

As $T(P_i)$ and $T^{-1}(P_i)$ are P_{i+1} measurable and $T(Q_j)$ and $T^{-1}(Q_j)$ are Q_{j+1} measurable, on this subset $X'\subset X$ of full measure

$$D(T(P_i)|Q_j)(T(x))\xrightarrow{j}D(P_i|\mathscr{A})(x)$$

and

$$D(T^{-1}(P_i)|Q_j)(T^{-1}(x)) \xrightarrow{j} D(P_i|\mathscr{A})(x)$$

For $x \in X'$, let $\mathscr{C}(x) = c_1(x) \supset c_2(x) \ldots$ be the chain of sets in the $\{P_i\}$ tree descending to x.

Let

$$\alpha(x) = \lim_{i \to \infty} E(c_i(x)|\mathscr{A}).$$

As $T(c_i(x))$ is a finite union of sets in P_{i+1}, one of which is $c_{i+1}(T(x))$, and similarly for T^{-1}, $\alpha(T(x)) = \alpha(x)$ for $x \in X'$ and α is $\mathscr{A}$ measurable. Thus there is a T-invariant subset $X'' \subset X'$ of full measure, and restricting to X'', α is constant on the equivalence classes of points not separated by the $\{Q_i\}$ tree.

The function $D(P_i|\mathscr{A})(x)$ is, for each $x \in X''$, a probability vector. The component terms of $D(P_{i+1}|\mathscr{A})(x)$ can be summed in subsets according to which element of P_i they belong, to obtain $D(P_i|\mathscr{A})(x)$.

It follows, that if $\alpha(x) > 0$, then for some $n \in \mathbb{Z}$, $\alpha(x) = 1/n$ and for some $I(x)$ large enough, if $i \geq I(x)$

$$D(P_i|\mathscr{A})(x) = \left\{\frac{1}{n}, \frac{1}{n}, \frac{1}{n}, \ldots, \frac{1}{n}\right\}, \tag{3.34}$$

where we have omitted elements of measure 0. To see (3.34) just notice that as i grows if arbitrarily small positive terms occur in $E(P_i|\mathscr{A})(x)$ then for some x' equivalent to x, $\alpha(x') = 0$. But $\alpha(x') = \alpha(x) > 0$. Hence, as of some stage $I(x)$, sets cease being split. As $\alpha(x)$ is a constant on the equivalence class, all non-zero terms in the vector must be equal.

Break X'' into T-invariant subsets $X_1, X_2, X_3, \ldots, X_\infty$ where $x \in X_n$ if $\alpha(x) = 1/n$ $(1/\infty = 0)$. We will discuss X_∞, as it is the more interesting case. The X_n, $n < \infty$ follow a similar line of reasoning leading to the parts of the decomposition of type 1. X_∞ is the part of type 2.

Assume $\mu(X_\infty) > 0$, and renormalizing μ, we can assume $X = X_\infty$ and for all x, $\alpha(x) = 0$.

We first construct a map $\bar{\varphi}$ from a.a. the equivalence classes of points not separated by $\{Q_i\}$ onto a.a. of $[0, 1]$ by assigning to successive levels of the tree left-closed, right-open intervals. We have discussed this construction in detail in Chapter 2. The map $\bar{\varphi}$ is defined a.e., maps to a.a. of $[0, 1]$ and is bimeasurable and measure-preserving. We can assume the subset X' where $\bar{\varphi}$ is defined is T-invariant. Let $Z = \bar{\varphi}(X')$. We wish to refine $\bar{\varphi}$ to a point map from X' to $Z \times [0, 1]$, still bimeasurable and measure-preserving by successively cutting the fibres $f_z = z \times [0, 1]$ according to the probability vectors $D(P_i|\mathscr{A})(\bar{\varphi}^{-1}(x))$.

Let $P_1 = \{p_1, p_2, \ldots, p_s\}$. The functions $D(P_i|\mathscr{A})$ are constant on $\{Q_i\}$ separated fibres, hence $\{D(P_i|\mathscr{A})(\bar{\varphi}^{-1}(z))\}_{i=1}^{s}$ is well-defined everywhere on X and is a measurable probability vector-valued function.

For a set $q_j \in Q_1$ and $p_i \in P_1$, define a map $\varphi(q_j \cap p_i)$ to the measurable subset of $\bar{\varphi}(q_j) \times [0, 1]$ between the graphs of

$$E\left(\bigcup_{k=1}^{i-1} p_i | \mathscr{A}\right) \circ \bar{\varphi}^{-1} \quad \text{and} \quad E\left(\bigcup_{k=1}^{i} p_i | \mathscr{A}\right) \circ \bar{\varphi}^{-1}, \tag{3.35}$$

closed below, open above. It follows easily that

$$\ell_2(\varphi(q_j \cap p_i)) = \mu(q_j \cap p_i) \tag{3.36}$$

where ℓ_2 is two-dimensional Lebesgue measure. Figure 3.2 illustrates this construction.

We extend φ to sets of the form

$$q_j \cap p_i, \quad q_j \in Q_k, \quad p_i \in P_k$$

by induction on k. We assume the elements of P_k are ordered so that those whose union is the first element of P_{k-1} come first, next those in the second element, etc. Thus defining $\varphi(q_j \cap p_i)$ to be the subset of $\bar{\varphi}(q_j) \times [0, 1]$ between the graphs of

$$E\left(\bigcup_{k=1}^{i-1} P_i | \mathscr{A}\right) \circ \bar{\varphi}^{-1} \quad \text{and} \quad E\left(\bigcup_{k=1}^{i} P_i | \mathscr{A}\right) \circ \bar{\varphi}^{-1},$$

closed below, open above, we automatically get

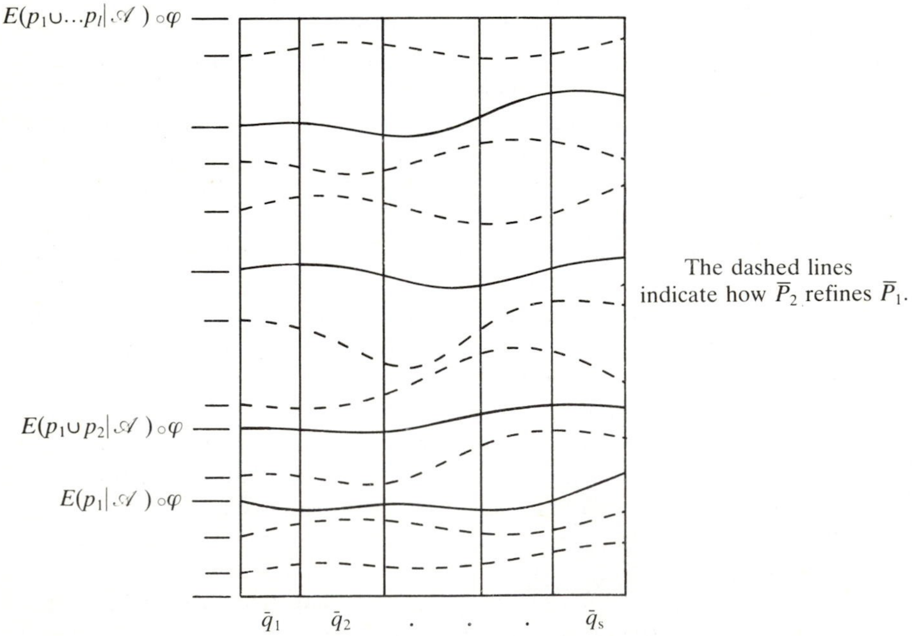

Fig. 3.2 Disintegration over a factor algebra. The dashed lines indicate how $\bar{P}_2$ refines $\bar{P}_1$.

1. $\ell_2(\varphi(q_j \cap p_i) = \mu(q_j \cap p_i)$.
2. $\{\varphi(q_j \cap p_i) | q_j \cap p_i$ is from level k of $\{Q_i \vee P_i\}\}$ partitions $Z \times [0, 1]$. Call this partition $\bar{Q}_i \vee \bar{P}_i$.
3. The partitions $\{\bar{Q}_i \vee \bar{P}_i\}$ form a tree of partitions exactly mirroring the intersection properties of $\{Q_i \vee P_i\}$.

The map φ gives a 1–1 correspondence between the chains of the $\{Q_i \vee P_i\}$ tree and the $\{\bar{Q}_i \vee \bar{P}_i\}$ tree of partitions. If two points $\mathbf{z}_1$ and $\mathbf{z}_2$ are not separated by the $\{\bar{Q}_i \vee \bar{P}_i\}$ tree then first, they must lie on the same fibre f_z, as the $\bar{Q}_i$ tree separates points of Z. But now, from our construction, the chain of sets $\bar{c}_1 \supset \bar{c}_2 \supset \cdots$ that contains $\mathbf{z}_1$ and $\mathbf{z}_2$ must intersect to an interval, and hence

$$E(\varphi^{-1}(\bar{c}_1)|\mathscr{A})(\bar{\varphi}^{-1}(\mathbf{z}_1)) \xrightarrow{i} \alpha > 0.$$

Now the $c_i = \varphi^{-1}(\bar{c}_i)$ either intersect to $\varnothing$ or to a single point $x \in X$. If not $\varnothing$, then $\alpha = \alpha(x) > 0$ which we know is not true. Hence the collection of all chains in $\{\bar{Q}_i \vee \bar{P}_i\}$ that descend to more than one point correspond to empty chains in $\{Q_i \vee P_i\}$. These form a set of chains of measure 0 in X, hence a set of measure 0 in $Z \times [0, 1]$. On what remains, $\{\bar{Q}_i \vee \bar{P}_i\}$ is a generating tree of partitions. The $\{\bar{Q}_i \vee \bar{P}_i\}$ chains that descend to $\varnothing$ form a set of chains of measure 0. Delete from X a T-invariant subset of measure 0 containing the intersection points of the chains corresponding to these.

On the remaining T-invariant set of full measure φ now reduces to a point map which is bimeasurable and measure-preserving as the two trees are images one of the other and generate.

Let $\bar{T} = \varphi T \varphi^{-1}$ be a measure-preserving map of $\varphi(X) \subset [0, 1] \times [0, 1]$ to itself. As $T(Q_1)$ and $T^{-1}(Q_i)$ are subsets of Q_{i+1}, $\bar{T}$ maps fibres f_z to fibres, but furthermore, as $T(P_i)$ and $T^{-1}(P_i)$ are subsets of P_{i+1}, $\bar{T}$ maps the intervals on a fibre corresponding to some P_k to a finite union of intervals on the image fibre in a measure-preserving fashion, the $\{\bar{P}_i\}$ tree restricted to a fibre generates the Lebesgue algebra $\mathscr{F}_z$ on it and so letting ℓ_z represent this fibre measure, $\bar{T}$ is a bimeasurable measure-preserving map from $(f(z), \mathscr{F}_z, \ell_z)$ to $(f(\bar{T}(z)), \mathscr{F}_{\bar{T}(z)}, \ell_{\bar{T}(z)})$.

As the $\bar{Q}_i$ consist of vertical sets and are $\varphi(Q_i)$, $\bar{\mathscr{A}} = \varphi(\mathscr{A})$ is the completion of the algebra of vertical sets. From Fubini's theorem, for any measurable set $A \subset [0, 1] \times [0, 1]$, for a.e. z, $A \cap f(z)$ is measurable

$$E(A|\bar{\mathscr{A}})(z) = \ell_z(A)$$

and

$$\ell(A) = \int_0^1 \ell_z(A)\, \mathrm{d}x.$$

For X_n, $n < \infty$, the construction of the map to almost all of $[0, 1] \times \{1, \ldots, n\}$

is completely analogous and as measurability on the fibres is trivial, is much easier. ■

Exercise 3.8 Suppose $\mu(X_n) > 0$. Show $X_n = \bigcup_{i=0}^{n-1} A_i$, a disjoint union, and $T^{-1}(A_i) = A_{i-1(\text{mod } n)}$. Note the similarity of this to Exercise 3.3 except here ergodicity of T is not assumed, just that $\mathscr{I} \subset \mathscr{A}$.

Corollary 3.18 *If in the above construction $\mathscr{A} = \mathscr{I} =$ the algebra of T-invariant* (mod 0) *sets, then for a.e. x, $\bar{T}$ is a bimeasurable measure-preserving ergodic map from $(F(z), \mathscr{F}_z, \ell_z)$ to itself. The disintegration of $(X, \mathscr{F}, \mu)$ over $\mathscr{I}$ is called the* ergodic decomposition *of the system.*

Proof After deleting a set of measure 0, $T(Q_i) = Q_i$ identically. Hence, a.s., $\bar{T}(f(z)) = f(z)$.

All that remains is to verify ergodicity. We use Corollary 3.16. Let f be the characteristic function of some set $p_k \in P_i$, and f_n its Cesáro averages. By the Birkhoff theorem, f_n converges a.e. to $E(f|\mathscr{I})$. As there are only countably many such f, for a.e. z, for ℓ_z a.e. y,

$$f_n((z, y)) \to E(f|\mathscr{I})(z),$$

a constant ℓ_z a.e. on $f(z)$. That this holds for finite unions of sets in some P_i follows easily. These are L^1 dense as $\{P_i\}$ generates, hence by Corollary 3.16 to the Birkhoff theorem $\bar{T}$ on $(f(z), \mathscr{F}_z, \ell_z)$ is ergodic for a.e. x. ■

This completes our discussion of disintegration over factor algebras. We will not argue the essential uniqueness of the disintegration. This is not terribly difficult to demonstrate.

4 Mixing properties

The fundamental problem of ergodic theory is to explore the structure of measure-preserving transformations in search of properties natural to them, which can be easily applied to describe and distinguish them. In this chapter we will discuss a hierarchy of such properties, each successively stronger, called mixing properties.

The reason for this name is that they concern the way in which the powers of a transformation T 'mix' one set, A, into another set B, i.e., they concern the sequence of functions

$$\chi_{T^{-i}(A)} \cdot \chi_B = \chi_{T^{-i}(A) \cap B}. \tag{4.1}$$

4.1 Poincaré recurrence

The simplest of these properties is Poincaré recurrence which says for some $i > 0$

$$\chi_{T^{-i}(A)} \cdot \chi_A \neq 0.$$

Theorem 4.1 *If T is a measure-preserving transformation of the probability space $(X, \mathscr{F}, \mu)$ and $\mu(A) > 0$, then for some $0 < i < [1/\mu(A)]$*

$$\mu(T^{-i}(A) \cap A) > 0. \tag{4.2}$$

Proof If $\mu(T^{-i}(A) \cap A) = 0$ for all such i, then

$$\mu(X) \geq \sum_{i=0}^{[1/\mu(A)]} \mu(T^{-i}(A)) = \mu(A)\left(1 + \left[\frac{1}{\mu(A)}\right]\right) > 1,$$

a conflict. ■

4.2 Ergodicity as a mixing property

If T is ergodic we know more. The L^2-ergodic theorem says

$$\frac{1}{n} \sum_{i=0}^{n-1} \int \chi_{T^{-i}(A)} \cdot \chi_B \, d\mu \xrightarrow{n} \mu(A)\mu(B)$$

which we could write as

$$\frac{1}{n}\sum_{i=0}^{n-1}\int(\chi_{T^{-i}(A)}\cdot\chi_B-\mu(A)\mu(B))\,d\mu\xrightarrow{n}0;$$

this is half of the following theorem.
Note: from here on 'transformation' will always mean an invertible, measure-preserving map from a Lebesgue probability space to itself.

Theorem 4.2 *A transformation T is ergodic iff for any two measurable sets A and B,*

$$\frac{1}{n}\sum_{i=0}^{n-1}\int(\chi_{T^{-i}(A)}\cdot\chi_B-\mu(A)\mu(B))\,d\mu\xrightarrow{n}0\quad\text{a.e.}\tag{4.3}$$

Proof We know one direction, that if T is ergodic the limit holds.
Assume the limit holds and suppose $T(A)=A$. Letting $A=B$

$$\frac{1}{n}\sum_{i=0}^{n-1}\int(\chi_{T^{-i}(A)}\cdot\chi_A-\mu(A)^2)\,d\mu\xrightarrow{n}0.$$

But as $T^{-1}(A)=A$ this says

$$\int\chi_A^2\,d\mu=\mu(A)=\mu(A)^2.$$

Thus $\mu(A)=0$ or 1 and T is ergodic. ∎

Corollary 4.3 *The transformation T is ergodic iff for any measurable set A*

$$\frac{1}{n}\sum_{i=0}^{n-1}\int(\chi_{T^{-i}(A)}\cdot\chi_A-\mu(A)^2)\,d\mu\xrightarrow{n}0.$$ ∎

For mixing properties this is not unusual, that knowing all sets mix with themselves in some fashion implies the same for any pair of sets.

4.3 Weakly mixing

Our first non-trivial mixing property will require that the Cesáro convergence of the sums above be absolute.

Definition 4.1 We say a transformation T is *weakly mixing* if for any measurable sets A and B,

$$\frac{1}{n}\sum_{i=0}^{n-1}\int|\chi_{T^{-i}(A)}\chi_B-\mu(A)\mu(B)|\,d\mu\xrightarrow{n}0.\tag{4.4}$$

This condition lies at the heart of a large web of argument. Perhaps the core result here is that any ergodic transformation has a maximal invariant factor algebra on which it is isomorphic to an isometry of a compact metric space. The transformation is weakly mixing exactly when this factor algebra is trivial. We will offer two proofs of this fact, one by a bare-hands construction of the metric space due to Katznelson, the other via a short discussion of spectral theory.

Our first result will already show that weakly mixing is a non-trivial condition. We first need a preliminary definition. We use the symbol # to indicate the cardinality of a finite set.

Definition 4.2 A subset $S \subset \mathbb{N}$ is of *density* α if

$$\frac{\#(S \cap \{0, 1, \ldots, n-1\})}{n} \xrightarrow{n} \alpha, \tag{4.5}$$

and of *full density* if $\alpha = 1$, i.e., S contains 'almost all' of $\mathbb{N}$.

Lemma 4.3 *A transformation T is weakly mixing iff for any measurable A and B there is a subset $S = \{n_1 < n_2 < n_3 \ldots\}$ of $\mathbb{N}$ of full density for which*

$$\lim_{k \to \infty} \mu(T^{-n_k}(A) \cap B) = \mu(A)\mu(B). \tag{4.6}$$

Proof Suppose such a subset S existed. Then

$$\overline{\lim_{k \to \infty}} \frac{1}{n} \sum_{i=0}^{n-1} \int |\chi_{T^{-i}(A)}\chi_B - \mu(A)\mu(B)| \, d\mu$$

$$\leq \overline{\lim_{n \to \infty}} \frac{1}{n}\left(\sum_{S \cap \{0, \ldots, n-1\}} |\mu(T^{-i}(A) \cap B) - \mu(A)\mu(B)|\right)$$

$$+ \overline{\lim_{n \to \infty}} \frac{1}{n}(\# S^c \cap \{0, 1, \ldots, n-1\})$$

which equals zero.

On the other hand, if T is weakly mixing,

$$\lim_{n \to \infty} \frac{1}{n} \sum_{i=0}^{n-1} |\mu(T^{-i}(A) \cap B) - \mu(A)\mu(B)| = 0.$$

Letting $S_\varepsilon = \{i : |\mu(T^{-i}(A) \cap B) - \mu(A)\mu(B)| < \varepsilon\}$,

$$\overline{\lim} \frac{\#\{S_\varepsilon^c \cap \{0, \ldots, n-1\}\}}{n} \leq \frac{1}{\varepsilon} \lim_{n \to \infty} \frac{1}{n} \sum_{i=0}^{n-1} |\mu(T^{-i}(A) \cap B) - \mu(A)\mu(B)| = 0.$$

Thus $\lim_{n \to \infty} \dfrac{\#\{S_\varepsilon \cap \{0, \ldots, n-1\}\}}{n} = 1$. If density in N were σ-additive, we could now just take $\bigcap_i S_{1/i}$. As it is not, we must be more clever. Choose $\{N_i\}$ so that $N_{i+1}/N_i > i$ and for $n \geq N_i$

$$\frac{\#\{S_{1/i} \cap \{0, \ldots, n-1\}\}}{n} > 1 - 1/i.$$

Let $S = \bigcup_{i=1}^{\infty} S_{1/i} \cap \{0, 1, \ldots, N_i - 1\}$. The result is now a computation. ■

Corollary 4.5 *T is weakly mixing iff*

$$\lim \frac{1}{n} \sum_{i=0}^{n-1} \int (\chi_{T^{-i}(A)}\chi_B - \mu(A)\mu(B))^2 \, d\mu = 0. \tag{4.7}$$

Theorem 4.6 *An ergodic isometry T of a compact metric space X is not weakly mixing. (We assume X contains more than one point.)*

Proof We may as well assume T is minimal as otherwise X decomposes into T-invariant closed sets on which it is and as T is ergodic, μ is supported on just one such set. Let $B_r(x)$ denote the open ball of radius r centred at x. Now $\mu(B_r(x))$ is independent of x and is a non-decreasing function of r and so is continuous at some point $r_0 > 0$. Thus for any $\varepsilon > 0$ there is a $\delta > 0$ so that if $d(x, y) < \delta$ then

$$\mu(B_{r_0}(x) \,\Delta\, B_{r_0}(y)) < \varepsilon\mu(B_{r_0}(x)).$$

Now $\mu(B_\delta(x)) > 0$, and the convergence of the Birkhoff theorem applied to $B_\delta(x)$ holds uniformly on X. (see Example 4 of Chapter 1 and Exercise 3.5 on unique ergodicity). Thus $\{n : T^n(x) \in B_\delta(x)\}$ has density $\mu(B_\delta(x)) > 0$. But for such an n,

$$\begin{aligned}
\mu(B_{r_0}(x) \cap T^{-n}(B_{r_0}(x)) &= \mu(B_{r_0}(x) \cap B_{r_0}(T^n(x)) \\
&\geq \mu(B_{r_0}(x)) - \varepsilon(\mu(B_{r_0}(x) \,\Delta\, B_{r_0}(T^n(x))) \\
&\geq (1 - \varepsilon)\mu(B_{r_0}(x))
\end{aligned}$$

and T is not weakly mixing. ■

Corollary 4.7 *If T, acting on $(X, \mathscr{F}, \mu)$, has a factor action measurably isomorphic to an isometry of a compact metric space then T is not weakly mixing.*

Proof If T is weakly mixing, then restricted to any factor it also is. ■

Exercise 4.1 Refine Theorem 4.6 to show that if T is a minimal isometry of a compact metric space X, μ its unique invariant Borel probability measure, then for any $F \in L^2(\mu)$ and $\varepsilon > 0$, show there is a sequence n_k of positive density with $\int |F(T^n(x))F(x) - \|F\|_2^2|^2 \, d\mu < \varepsilon$.

We will now develop a circle of equivalent definitions of weakly mixing, one piece of which will be the converse of Corollary 4.7.

Theorem 4.8 *Let T acting on $(X, \mathscr{F}, \mu)$ be ergodic. If the Cartesian square $T \times T$, acting on $(X \times X, \mathscr{F} \times \mathscr{F}, \mu \times \mu)$ is ergodic then T is weakly mixing.*

The proof rests on the following piece of arithmetic.

Lemma 4.9 *If a_n is a sequence of real numbers with*

$$\lim_{n\to\infty} \frac{1}{n} \sum_{i=0}^{n-1} a_i = a \tag{4.8}$$

and

$$\lim_{n\to\infty} \frac{1}{n} \sum_{i=0}^{n-1} a_i^2 = a^2$$

then

$$\lim_{n\to\infty} \frac{1}{n} \sum_{i=0}^{n-1} (a_i - a)^2 = 0.$$

Proof

$$\lim_{n\to\infty} \frac{1}{n} \sum_{i=0}^{n-1} (a_i - a)^2 = \lim_{n\to\infty} \left(\frac{1}{n} \sum_{i=0}^{n-1} a_i^2 - \frac{2a}{n} \sum_{i=0}^{n-1} a_i + a^2 \right) = 0. \qquad \blacksquare$$

Proof of Theorem 4.7 For any sets A and B,

$$\lim_{n\to\infty} \frac{1}{n} \sum_{i=0}^{n-1} \mu(A \cap T^{-i}(B)) = \mu(A)\mu(B)$$

as T is ergodic.

Letting $\bar{A} = A \times A$ and $\bar{B} = B \times B$,

$$\begin{aligned} \lim_{n\to\infty} \frac{1}{n} \sum_{i=0}^{n-1} \mu \times \mu(\bar{A} \cap T^{-i} \times T^{-i}(\bar{B})) &= \lim_{n\to\infty} \frac{1}{n} \sum_{i=0}^{n-1} \mu(A \cap T^{-i}(B))^2 \\ &= \mu \times \mu(\bar{A})\mu \times \mu(\bar{B}) \\ &= (\mu(A)\mu(B))^2. \end{aligned}$$

By Lemma 4.9 then

$$\lim_{n\to\infty} \frac{1}{n} \sum_{i=0}^{n-1} (\mu(A \cap T^{-i}(B)) - \mu(A)\mu(B))^2 = 0$$

and Corollary 4.5 tells us T is weakly mixing. $\blacksquare$

The next theorem is Katznelson's demonstration that if a transformation is not weakly mixing, then it must have a factor isomorphic to a minimal isometry. This is done by constructing an invariant pseudometric which makes the space precompact. Later we will give an alternate proof via spectral theory.

Theorem 4.10 *If T acting on $(X, \mathscr{F}, \mu)$ is ergodic and has no factor actions isomorphic to an isometry of a compact metric space and S acting on $(Y, \mathscr{G}, \nu)$ any other ergodic transformation, then $T \times S$ acting on $(X \times Y, \mathscr{F} \times \mathscr{G}, \mu \times \nu)$ is ergodic.*

Proof We verify the contrapositive, i.e., assuming the Cartesian product is non-ergodic we construct a factor algebra on which T is isomorphic to an isometry of a compact metric space. We do this by constructing a non-trivial T-invariant pseudometric on X making it a precompact metric space. Let $A \in \mathscr{F} \times \mathscr{G}$ be a $T \times S$-invariant set, $0 < \mu \times \nu(A) = \alpha < 1$.

Let $\mathscr{F}_1 = \mathscr{F} \times \{\text{trivial algebra}\}$ and $\mathscr{F}_2 = \{\text{trivial algebra}\} \times \mathscr{G}$. Both are $T \times S$-invariant algebras, and on each, $T \times S$ is ergodic. Thus $E(\chi_A | \mathscr{F}_1)$ is $\mathscr{F}_1$ measurable and $T \times S$-invariant, hence is α a.s.

Letting $A_x = \{y | (x, y) \in A\}$, for a.e. x, A_x is measurable and $\nu(A_x) = E(\chi_A | \mathscr{F}_1) = \alpha$, and $A_{T(x)} = S(A_x)$ as A is $T \times S$-invariant.

Let $X_0 \subset X$ be a T-invariant set of full measure with $\nu(A_x) = \mu \times \nu(A) = \alpha$ for $x \in X_0$. For $x, x' \in X_0$, let $d(x, x') = \nu(A_x \Delta A_{x'})$.
Now

$$d(T(x), T(x')) = \nu(A_{T(x)} \Delta A_{T(x')}) = \nu(S(A_x \Delta A_{x'})) = d(x, x'),$$

and as

$$0 \leq d(x, x'') \leq d(x, x') + d(x', x''),$$

d is a T-invariant pseudometric, and hence a metric on the equivalence classes $\langle x \rangle = \{x^1 : d(x, x^1) = 0\}$.

Let $E = \{\langle x \rangle\}$, the space of such equivalence classes. As T is ergodic and $\mu \times \nu(A) \neq 0, 1$, E contains more than one point. For any set $B \in \mathscr{F}$, the function $f_B : x \to \mu(A_x \cap B)$ is constant on all x in a class $\langle x \rangle \in E$, and such functions separate the classes in E. Hence $(E, \mathscr{B}, \eta)$ is a measurable factor of $(X, \mathscr{F}, \mu)$.

Set $\phi : x \to \langle x \rangle$ to be the factor map. The measure μ, of course, projects to a measure η. We want to show that $U = T\phi$ acting on $(E, \mathscr{B}, \eta)$ is measurably isomorphic to an isometry of a compact metric space. It is sufficient for this to show that E contains a U-invariant subset E_0, of full measure, on which d is precompact. The map U will extend to an isometry on the compactification of E_0. Ergodicity of T implies U has a dense orbit, and hence is minimal and uniquely ergodic. μ must then project to this unique invariant probability measure, and E_0 is almost all of the compactification. Showing the existence of a precompact E_0 is the same as showing, for any $\varepsilon > 0$, X can be covered a.s. by a finite number of ε-balls in the pseudometric d.

Let $\{A_1, A_2, \ldots\}$ be a sequence of sets dense in $\mathscr{G}$, i.e., for any $A \in \mathscr{G}$ and $\varepsilon > 0$ there is an A_i with $\nu(A \,\Delta\, A_i) < \varepsilon$.

For ε fixed, let

$$B_i = \left\{x : \nu(A_i \,\Delta\, A_x) < \frac{\varepsilon}{6}\right\}$$

and

$$\bigcup_{i=1}^{\infty} B_i = X_0.$$

If $z \in T^j(B_i)$ then $\nu(A_z \,\Delta\, S^j(A_i)) = \nu(A_{S^{-j}(z)} \,\Delta\, A_i) < \varepsilon/6$.

Without loss of generality assume $\mu(B_1) > 0$ and now select N so large that

$$\mu\left(\bigcup_{i=1}^{N} B_i\right) > 1 - \frac{\mu(B_1)}{2}.$$

Let $X_1 = \bigcup_{i=-\infty}^{\infty} T^i(B_1)$, a set of full measure by ergodicity of T. For $x \in X_1$, $x \in T^j(B_1)$, as $\mu(T^j(B_1)) = \mu(B_1) > 1 - \mu(\bigcup_{i=1}^{N} B_i)$, there must be a $y \in T^j(B_1) \cap B_k$ for some $k \in \{1, \ldots, N\}$. But then

$$\nu(A_x \,\Delta\, A_k) \leq \nu(A_x \,\Delta\, T^j(A_i)) + \nu(T^i(A_1) \,\Delta\, A_y) + \nu(A_y \,\Delta\, A_k) < \frac{\varepsilon}{2},$$

and so the sets

$$A_i = \left\{x : \nu(A_x \,\Delta\, A_i) < \frac{\varepsilon}{2}\right\}, \quad i = 1, \ldots, N$$

cover X_1 and are of diameter less than ε, completing the result. ■

4.4 A little spectral theory

Having just completed one proof that a transformation is weakly mixing if and only if it has no non-trivial isometric factors, we now present another. In fact, our true intention is to show that a transformation is weakly mixing if and only if it has no non-trivial eigenfunctions. We do this by developing a small piece of the spectral theory of transformations. This material is self-contained and can be omitted or read lightly at first passage. It provides, however, an irreplaceable tool in ergodic theory.

We regard a transformation T as a unitary operator on complex-valued $L^2(\mu)$. What we want to do is to model this operator by multiplication by $e^{2\pi i\theta}$ on L^2 of the unit circle S^1 in the complex plane. We must make two sacrifices in order to do this. First we will not model all of $L^2(\mu)$ but only the T-invariant subspace generated by some single function. Second, on the unit circle, we will not have Lebesgue measure. In fact, the core work here is the construction of the appropriate Borel measure on S^1.

This second issue is not in fact a sacrifice. The measure we build becomes a kind of bookkeeper for much of the structure of T.

To begin, let T be an ergodic transformation on $(X, \mathscr{F}, \mu)$ and $F: X \to \mathbb{C}$ a complex-valued function in $L^2(\mu)$. We will define a spectral measure associated to F, (sp(f)). On $L^2(\text{sp}(f))$ we have a unitary operator, multiplication by the function $e^{2\pi i\theta}$. We want this operator to be isomorphic to the action of T on the subspace of $L^2(\mu)$ generated by F. In this correspondence F is to be associated to the function 1.

We will construct sp(F) by describing inner products of continuous functions with respect to sp(F). We need a standard way of uniformly approximating a continuous function by trigonometric polynomials. In our work we represent S^1 as $\{e^{i\theta}: 0 \le \theta \le 2\pi\}$, and write functions of S^1 as 2π-periodic functions of $\theta \in \mathbb{R}$. Let

$$K_n(\theta) = \sum_{j=-n}^{n} \left(1 - \frac{|j|}{n+1}\right) e^{ij\theta} = \frac{1}{n+1} \left(\frac{\sin\left(\frac{n+1}{2}t\right)}{\sin \frac{1}{2} t} \right)^2 \tag{4.9}$$

be Fejer's kernel (Katznelson 1968).

Lemma 4.10 *Fejer's kernel is a positive summability kernel in that*

(1) $K_n(\theta)$ *is* 2π*-periodic, continuous and non-negative for all* n, θ;

(2) $1/2\pi \int K_n(\theta)\,d\theta = 1$; *and*

(3) $\lim_{n\to\infty} K_n(\theta) = 0$ *uniformly on any interval* $\delta \le \theta \le 2\pi - \delta, 0 < \delta < \pi$.

Proof Exercise 4.2 part (1).

Corollary 4.12 *For any continuous* 2π*-periodic* $f: \mathbb{R} \to \mathbb{C}$,

$$\sigma_n(f)(\theta) = \frac{1}{2\pi} \int_0^{2\pi} f(t-\theta) K_n(t)\,dt = \sum_{j=-n}^{n} c_j^n(f) e^{ij\theta} \tag{4.10}$$

is a trigonometric polynomial

$$c_j^n(f) = \left(1 - \frac{|j|}{n+1}\right) \frac{1}{2\pi} \int_0^{2\pi} f(t) e^{ijt}\,dt = \left(1 - \frac{|j|}{n+1}\right) c_j(f). \tag{4.11}$$

Further, σ_n *is a positive linear operator*

$$\|f\|_\infty \ge \|\sigma_n(f)\|_\infty$$

and $\sigma_n(f) \xrightarrow[n]{} f$ *uniformly.*

Proof Exercise 4.2 part (2).

Exercise 4.2

(1) Prove Lemma 4.11.

(2) Prove Corollary 4.12.

Let f and g be two 2π-periodic continuous functions from $\mathbb{R}$ to $\mathbb{C}$. Define a series of bilinear forms

$$\begin{aligned}\langle f, g\rangle_n &= \sum_{j,k=-n}^{n} c_j^n(f)\overline{c_k^n}(g)a_{j,k} \\ &= \int\left(\sum_{j=-n}^{n} c_j^n(f)F(T^j(x))\right)\left(\overline{\sum_{k=-n}^{n} c_k^n(g)F(T^k(x))}\right)\mathrm{d}\mu. \qquad (4.12)\end{aligned}$$

We want to see that these bilinear forms converge to the desired inner product.

The next result is a special case of Bochner's theorem.

Theorem 4.13 *Suppose f is continuous, 2π-periodic, real-valued and positive. The operators $f \to \langle\sqrt{f}, \sqrt{f}\rangle_n$ are in fact integrals with respect to real positive Borel measures ν_n on $[0, 2\pi)$, each of total mass $a_{0,0}$.*

These measures ν_n converge weak to a measure we call* sp(F) *which, for continuous f and g, satisfies*

$$\langle f, g\rangle_n = \langle \sigma_n(f), \sigma_n(g)\rangle_{\mathrm{sp}(n)}.$$

Proof For f in the cone of positive continuous functions $L_n : f \to \langle\sqrt{f}, \sqrt{f}\rangle_n$ is continuous in the uniform norm on f. We can extend L_n to all of $C_{\mathbb{R}}([0, 2\pi))$ by $L(f^+) - L(f^-)$ and get a complex-valued continuous linear functional.

Hence by the Riesz representation theorem, $L_n(f) = \int f\,\mathrm{d}\nu_n$ for some perhaps complex-valued measure ν_n.

But for $f \geq 0$ we compute

$$\int f^2\,\mathrm{d}\nu_n = \langle f, f\rangle_n = \int\left|\sum_{k=-n}^{n} c_k^n(f)F(T^k(x))\right|^2 \mathrm{d}\mu,$$

so ν_n is a positive real-valued measure with

$$\nu_n([0, 2\pi)) = \langle 1, 1\rangle_n = a_{0,0}.$$

The ν_n lie in a bounded, hence weak*, compact region of $C_{\mathbb{R}}^*([0, 2\pi))$. If $P = \sum_{j=-N}^{N} c_j \mathrm{e}^{ijt}$ is a trigonometric polynomial

$$\int |P|^2\,\mathrm{d}\nu_n \underset{n}{\to} \sum_{n=-N}^{N} c_j\bar{c}_k a_{j,k}.$$

Any positive continuous f is the uniform limit of such polynomials $|P|^2$. Thus the sequence of measures $\{\nu_n\}$ cannot have more than one weak* limit and so must converge weak* to a Borel measure we call sp(F).

We want to identify

$$\int fg \, \mathrm{d}(\mathrm{sp}(F)) = \langle f, g \rangle_{\mathrm{sp}(F)}.$$

Let P and Q be positive real-valued trigonometric polynomials

$$P = \sum_{j=-N}^{N} b_j \mathrm{e}^{\mathrm{i}j\theta}$$

$$Q = \sum_{k=-N}^{N} c_k \mathrm{e}^{\mathrm{i}k\theta}.$$

Using

$$\begin{aligned} 2\langle P, Q \rangle_{\mathrm{sp}(F)} &= \langle P, Q \rangle_{\mathrm{sp}(F)} + \langle P, Q \rangle_{\mathrm{sp}(F)} \\ &= (\langle P + Q, P + Q \rangle_{\mathrm{sp}(F)} - \langle P, P \rangle_{\mathrm{sp}(F)} \langle Q, Q \rangle_{\mathrm{sp}(F)}) \end{aligned}$$

one easily computes

$$\langle P, Q \rangle_{\mathrm{sp}(F)} = \sum_{j,k=-N}^{N} b_j \bar{c}_k a_{j,k}.$$

This easily extends by linearity to arbitrary complex-valued polynomials. Thus for f and g continuous

$$\langle \sigma_n(f), \sigma_n(g) \rangle_{\mathrm{sp}(F)} = \sum_{j,k=-N}^{N} c_j^n(f) \bar{c}_k^n(g) a_{j,k} = \langle f, g \rangle_n. \tag{4.13}$$

As $\sigma_n(f)$ and $\sigma_n(g)$ converge uniformly to f and g,

$$\langle f, g \rangle_{\mathrm{sp}(F)} = \lim_{n \to \infty} \langle \sigma_n(f), \sigma_n(g) \rangle_{\mathrm{sp}(F)} = \lim_{n \to \infty} \langle f, g \rangle_n. \qquad \blacksquare$$

Define now a map Φ_F : {trigonometric polynomials} $\to L^2(\mu)$ by setting

$$\Phi_F\left(\sum_{j=-N}^{N} b_j \mathrm{e}^{\mathrm{i}j\theta} \right) = \sum_{j=-N}^{N} b_j F(T^j(x)). \tag{4.14}$$

Thus, for example, we can compute

$$\Phi_F(1) = F(x)$$

and

$$\Phi_F(\mathrm{e}^{\mathrm{i}\theta} P(\theta)) = \Phi_F(P(\theta)) \circ T$$

for any trigonometric polynomial P.

The next result is the core of spectral theory as we will use it. Let $L^2(F, \mu)$ be the closure in $L^2(\mu)$ of the linear span of the set of functions $\{F \circ T^j\}_{j \in \mathbb{Z}}$.

Theorem 4.14 *The map Φ_F extends to an L^2-isometry from $L^2(\mathrm{sp}(F))$ to $L^2(F, \mu)$. As we have seen, $\Phi_F(1) = F$ and for any $f \in L^2(\mathrm{sp}(F))$,*

$$\Phi_F(e^{i\theta} f(\theta)) = \Phi(f) \circ T.$$

Proof For P and Q polynomials,

$$\begin{aligned}\langle P, Q \rangle_{\mathrm{sp}(F)} &= \sum_{j,k} b_j \bar{c}_k a_{j,k} \\ &= \int \Big(\sum_j b_j F(T^j(x))\Big)\Big(\sum_k c_k F(T^k(x))\Big)\, d\mu \\ &= \langle \Phi_E(P), \Phi_F(Q) \rangle_\mu.\end{aligned}$$

Thus Φ_F is an L^2-isometry where it is defined. Certainly then it extends isometrically to the closure of the polynomials in $L^2(\mathrm{sp}(F))$. This is all of $L^2(\mathrm{sp}(F))$.

The image under Φ_F of the trigonometric polynomials is exactly the linear span of $\{F \circ T^j\}_{j \in \mathbb{Z}}$. Hence the range of the extended Φ_F is its closure.

Since T acts as an isometry of $L^2(F, \mu)$ to itself, as does multiplication by $e^{i\theta}$ on $L^2(\mathrm{sp}(F))$, the identity

$$\Phi_F(e^{i\theta} f(\theta)) = \Phi_F(f) \circ T$$

extends from polynomials to all of $L^2(\mathrm{sp}(F))$. ∎

Corollary 4.15 *If $G \in L^2(F, \mu)$, then $\mathrm{sp}(G) \prec \mathrm{sp}(F)$ and we can compute the Radon–Nikodym derivative*

$$\frac{d\mathrm{sp}(G)}{d\mathrm{sp}(F)} = |\Phi_F^{-1}(G)|^2.$$

Proof Supposing $G = \Phi(P)$, P a polynomial and f and g are also polynomials, it is a finite computation that

$$\langle Pf, Pg \rangle_{\mathrm{sp}(F)} = \langle f, g \rangle_{\mathrm{sp}(\Phi_F(P))}.$$

Suppose now $G \in L^2(F, \mu)$ is arbitrary, $G = \Phi_F(h)$. Suppose $P_i \underset{i}{\rightarrow} h$ in $L^2(\mathrm{sp}(F))$, where the P_i are polynomials.

Then $G_i = \Phi_F(P_i) \to G$ in $L^2(\mu)$. Thus all the coefficients

$$\int G_i(T^j(x))\overline{G_i(T^k(x))}\, d\mu \underset{i}{\rightarrow} \int G_i(T^j(x)) G_i(T^k(x))\, d\mu.$$

This tells us, for f and g still polynomials

$$\langle hf, hg \rangle_{\mathrm{sp}(F)} = \langle f, g \rangle_{\mathrm{sp}(G)}.$$

As Φ_F and Φ_G are both L^2 isometries, this tells us that the map $f \to gf$ is an L^2-isometry from $L^2(\mathrm{sp}(G))$ into the ideal $hL^2(\mathrm{sp}(F))$, and for any $f \in$

$L^2(\mathrm{sp}(G))$,

$$\int f\,\mathrm{dsp}(G) = \int |h|^2 f\,\mathrm{dsp}(F).$$ ■

A complete discussion of the spectral theory of ergodic transformations would now include how $L^2(\mu)$ is built up from the pieces $L^2(F,\mu)$. We have what we want of spectral theory now, though, and refer the reader to Parry (1969) to pursue this picture further.

4.5 Weakly mixing and eigenfunctions

We will now see how a spectral measure $\mathrm{sp}(F)$ can ferret out the failure of weakly mixing.

Lemma 4.16 *For $F \in L^2(\mu)$,*

$$\frac{1}{2n+1}\sum_{j=-n}^{n}\left|\int F(x)\overline{F(T^j(x))}\,\mathrm{d}\mu\right|$$

tends to zero in n iff $\mathrm{sp}(F)$ *has no atoms.*

Proof To identify the atoms of $\mathrm{sp}(F)$ we look at $[0,2\pi)\times[0,2\pi)$ with measure $\nu = \mathrm{sp}(F)\times\mathrm{sp}(F)$. Notice that

$$\nu(\{(\theta,\theta):\theta\in[0,2\pi)\}) = (\mathrm{sp}(F)(\text{atomic part}))^2.$$

To compute $\nu(\{(\theta,\theta)\})$ consider the sequence of functions

$$f_n(x,y) = \frac{1}{2n+1}\sum_{j=-n}^{n} \mathrm{e}^{\mathrm{i}j(x-y)} \tag{4.15}$$

which converge pointwise to the characteristic function of the diagonal and are uniformly bounded by 1. Thus

$$\nu(\{(\theta,\theta)\}) = \lim_{n\to\infty}\int f_n(x,y)\,\mathrm{d}\nu.$$

Computing,

$$\begin{aligned}\int f_n(x,y)\,\mathrm{d}\nu &= \frac{1}{2n+1}\sum_{j=-n}^{n}\int \mathrm{e}^{\mathrm{i}jx}\mathrm{e}^{-\mathrm{i}jy}\,\mathrm{d}\nu\\ &= \frac{1}{2n+1}\sum_{j=-n}^{n} a_{0,j}a_{0,-j}\\ &= \frac{1}{2n+1}\sum_{j=-n}^{n}\left|\int F(x)\bar{F}(T^j(x))\,\mathrm{d}\mu\right|^2.\end{aligned}$$ ■

Corollary 4.17 *Suppose T is ergodic. For any set $A \in \mathscr{F}$,*

$$\frac{1}{2n+1}\sum_{j=-n}^{n}(\mu(A \cap T^{-j}(A)) - \mu(A)^2)^2 \tag{4.16}$$

tend to zero in n iff $\mathrm{sp}(F)$ *has no atoms away from* 0, *where*

$$F = \chi_A(x) - \mu(A).$$

Proof $F \in L^2(\mu)$ and notice

$$\begin{aligned}\frac{1}{2n+1}\sum_{j=-n}^{n}&\left|\int F(x)\overline{F(T^j(x))}\,\mathrm{d}\mu\right|^2\\ &= \frac{1}{2n+1}\sum_{j=-n}^{n}\left(\int \chi_A(x) - \mu(a))(\chi_A(T^j(x)) - \mu(A))\right)^2\\ &= \frac{1}{2n+1}\sum_{j=-n}^{n}(\mu(A \cap T^j(A)) - \mu(A)^2)^2.\end{aligned}$$

Furthermore,

$$\mathrm{sp}(F)(\{0\}) = \lim_{n\to\infty}\frac{1}{2n+1}\sum_{j=-n}^{n}\int F(x)\bar{F}(T^j(x))\,\mathrm{d}\mu = \int F\,\mathrm{d}\mu\int \bar{F}\,\mathrm{d}\mu$$

by the L^2-ergodic theorem (Theorem 3.1).

This is zero as $\int F\,\mathrm{d}\mu = 0$. Thus $\mathrm{sp}(F)$ has no atom at 0. By Lemma 4.16 it has no atoms elsewhere if and only if

$$\frac{1}{2n+1}\sum_{j=-n}^{n}(\mu(A \cap T^j(A)) - \mu(A)^2)^2$$

tends to zero. ∎

Notice that

$$\begin{aligned}\frac{1}{2n+1}&\sum_{j=-n}^{n}(\mu(A \cap T^j(A)) - \mu(A)^2)^2\\ &= \left(\frac{2n+2}{2n+1}\right)\left(\frac{1}{n+1}\sum_{j=0}^{n}(\mu(A \cap T^j(A)) - \mu(A)^2)^2\right) - \frac{(\mu(A) - \mu(A)^2)^2}{2n+1}.\end{aligned}$$

Thus the limit of the symmetric average is the same as that of the one-sided average.

Corollary 4.18 *Suppose T is ergodic but not weakly mixing. There is then a $\lambda \in \mathbb{C}$, $|\lambda| = 1$, $\lambda \neq 1$ and an $f \in L^2(\mu)$, $|f| = 1$ with $f(T(x)) = \lambda f(x)$, i.e., T has an eigenfunction with eigenvalue λ.*

Proof As T is not weakly mixing for some non-trivial set A and $F(x) = \chi_A(x) - \mu(A)$, we know $\mathrm{sp}(F)$ has an atom not at 0. Suppose it is at a point θ_0. Set $\lambda = \mathrm{e}^{\mathrm{i}\theta_0}$.

The function

$$\delta_{\theta_0} = \begin{cases} 1 & \text{at } \theta_0 \\ 0 & \text{elsewhere} \end{cases}$$

is in $L^2(\mathrm{sp}(F))$ and

$$\int \delta_{\theta_0}\, \mathrm{dsp}(F) > 0.$$

Thus $f = \Phi_F(\delta_{\theta_0})$ is not identically zero.

But notice

$$f(T(x)) = \Phi_F(\mathrm{e}^{\mathrm{i}\theta}\delta_{\theta_0}(\theta)) = \Phi(\lambda\delta_{\theta_0}) = \lambda f(x), \quad \mu\text{–a.s.}$$

To see that $f \in L^\infty(\mu)$ just note

$$|f(T(x))| = |\lambda||f(x)| = |f(x)|$$

and so $|f|$ is a constant μ–a.s. As $f \not\equiv 0$, we can normalize f to have modulus 1. ■

The eigenfunction f above can be regarded as a factor map from X to S^1 carrying the action of T to rotation by θ_0.

If θ_0 is irrational we know R_{θ_0} is uniquely ergodic, hence $f(\mu)$ must be Lebesgue measure on S^1. If θ_0 is rational then R_{θ_0} is not ergodic, but is still uniquely ergodic on each of its ergodic components. They are finite sets of course. Thus $f(\mu)$ must map to exactly one of them. In either case, we have obtained an isometric factor, giving an alternate proof of Theorem 4.10.

We have in fact gained more, having identified the isometry as either a rotation on the circle or on a finite point set. Thus, for example, we have shown that any isometry of a compact metric space must have such a factor algebra.

As at the end of Theorem 4.10, we are once more poised ready to show that an ergodic T has a maximal isometric factor. To do so along the lines of the proof of Theorem 4.10 involves constructing an appropriate metric from the full algebra of invariant sets in $X \times X$. Here, using spectral theory, we can more explicitly describe this algebra as the algebra generated by the eigenfunctions. The full discussion of this we construct as a series of exercises.

Exercise 4.3

1. Using Exercise 4.1, show that if T is a minimal isometry of a compact metric space and $F \in L^2(\mu)$ then $\mathrm{sp}(F)$ must have atoms.
2. Using Corollary 4.15 show that in fact $\mathrm{sp}(F)$ must be purely atomic. Hint: pick G to avoid atoms.

3. Conclude that for a minimal isometry, $L^2(\mu)$ is generated by eigenfunctions.
4. For T arbitrary but ergodic, let $\mathscr{A}$ be the algebra generated by the eigenfunctions. Show $\mathscr{A}$ must contain all isometric factors.
5. Let $\Lambda(T)$ be the set of all eigenvalues of T. Show that, as T is ergodic and $L^2(\mu)$ is separable, that $\Lambda(T)$ is countable.
6. For each $\lambda \in \Lambda(T)$, let $S_\lambda \subseteq L^2(\mu)$ be the subspace of eigenfunctions with eigenvalue λ. Show that as T is ergodic, $\dim(S_\lambda) = 1$.
7. Let $\Lambda(T) = \{\lambda_i\}$ and $f_i \in S_{\lambda_1}$ be a generator for S_{λ_1}. Show that the pseudometric $d(x, y) = \sum_i |f_i(x) - f_i(y)|/2^i$ makes the action of T on the factor algebra $\mathscr{A}$ isomorphic to a minimal isometry of a compact metric space.

This completes the argument that $\mathscr{A}$ is the maximal isometric factor, it gives more, showing that T is an isometry exactly when all $\mathrm{sp}(F)$ are purely atomic. A minimal isometry is in fact determined, up to isomorphism, by its eigenvalues $\Lambda(T)$. A proof of this result of von Neumann is accessible from our current vantage. It follows from the observation that if the eigenvalues agree, then one can construct an isometry between the two metrics as given in Exercise 4.3, part 7, above. This requires a careful analysis of how the various eigenfunctions must fit together.

Thus spectral theory, when the spectral measures are purely atomic is carried in the collection of numbers $\Lambda(T)$, and is relatively simple. When the spectral measures $\mathrm{sp}(F)$ are non-atomic the situation is much more difficult. We will stop here though, and return to tie together all our discussions of the weakly mixing property.

Proposition 4.19 *For an ergodic transformation T acting on $(X, \mathscr{F}, \mu)$, the following are equivalent*:

(1) *T is weakly mixing*;

(2) *the Cartesian product of T with any other ergodic transformation, with product measure, is ergodic*;

(3) *the Cartesian square $T \times T$ with product measure, is ergodic*;

(4) *T has no factor action isomorphic to an isometry of a compact metric space*;

(5) *T has no non-trivial eigenfunctions.* ∎

This circle of equivalent conditions makes the weakly mixing property extremely easy to work with. In Chapter 6 we will add another condition equivalent to these three, that of disjointness from all isometries. For a continuation of these investigations into the weakly mixing property, see Furstenberg (1981).

4.6 Mixing

Definition 4.3 Our next mixing property of a transformation T, simply called *mixing*, is that for any two sets $A, B \in \mathscr{F}$

$$\lim_{n\to\infty} \mu(T^n(A) \cap B) = \mu(A)\mu(B). \tag{4.17}$$

Thus the limit on a set of full density of Lemma 4.4 is made a limit.

This would seem to be the most natural of mixing conditions, having a far simpler definition than weak mixing (or the later K-mixing). The fact that we will prove no theorems about mixing is one indication that this is not the case. The mixing property is in fact rather difficult to work with.

Definition 4.4 We say T is *k-fold mixing* if for any sets $A_1, A_2, \ldots, A_k$

$$\lim_{n_2,\ldots,n_k\to\infty} \mu(A_1 \cap T^{n_2}(A_2 \cap T^{n_3}(A_3 \ldots (A_{k-1} \cap T^{n_k}(A_k)) \ldots)$$
$$= \mu(A_1)\mu(A_2)\ldots\mu(A_k). \tag{4.18}$$

There is a parallel notion of k-fold weak mixing, that the full density version of this limit holds. Furstenberg (1981) has shown that weakly mixing implies k-fold weakly mixing for all k. The relationship between two-fold and three-fold mixing remains one of the outstanding unsolved problems in measurable dynamics.

Chacón's transformation

We will now describe the construction of a transformation T, due to Chacón, which is weakly mixing but not mixing. We give a rank-1 cutting and stacking version of the construction. As described in Example 5 of Chapter 1, for Chacón's example

$$k(n) = 3, \quad S(1,n) = S(3,n) = 0, \quad S(2,n) = 1,$$

i.e., at each stage of the construction the stack is cut into three equal slices, which are stacked in order with a single new level placed between the second and third slices (Fig. 4.1). It is a computation that T acts on the interval $[0,3/2)$.

Theorem 4.20 *Chacón's map is weakly mixing but not mixing.*

Proof Let $N(n)$ be the number of intervals in the nth stack, and label these intervals

$$I(1,n), I(2,n), \ldots, I(N(n),n)$$

in order from the bottom of the stack upward. Thus for $1 \leq j < N(n)$, $T(I(j,n)) = I(j+1,n)$.

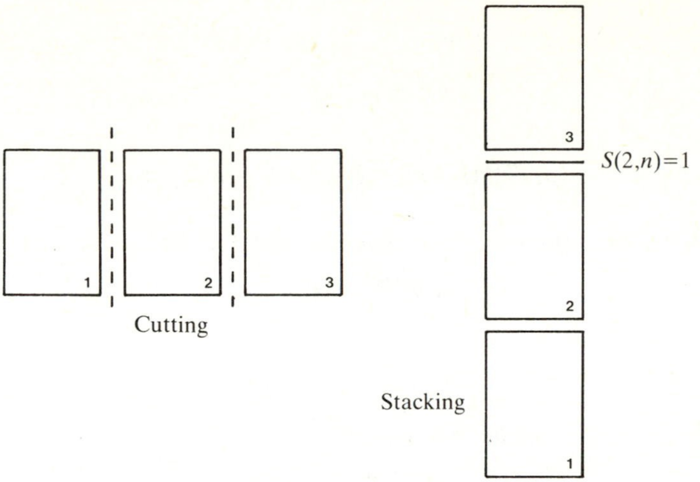

Fig. 4.1 Cutting and stacking in Chacon's map.

From the diagram above, for $1 < j < N(n)$,

$$\mu(T^{N(n)}(I(j,n)) \cap I(j,n)) \geq \frac{1}{3}\mu(I(j,n))$$

$$\mu(T^{N(n)+1}(I(j,n)) \cap I(j,n)) \geq \frac{1}{3}\mu(I(j,n)). \tag{4.19}$$

Let $S = I(2,2)$. For the second level of the second stack, $\mu(S) = 2/9$. For any $n > 2$, S is a disjoint union of sets of the form $I(j,n)$ where $1 < j < N(n)$. It follows that for $n > 2$,

$$\mu(T^{N(n)}(S) \cap S) \geq \frac{1}{3}\mu(S) > \mu(S)^2 \tag{4.20}$$

and T is not mixing.

We show T is weakly mixing by contradiction. Suppose $d(\ ,\ )$ is a non-trivial T-invariant pseudometric on $[0,3/2]$ making it precompact.

For $1/10 > \varepsilon > 0$, let D be an ε-ball of positive measure $\neq 1$. As the intervals $I(j,n)$ refine in n, there must be an n and j with $I(j,n)$ satisfying

$$\mu(I(j,n) \cap D) \geq (1-\varepsilon)\mu(I(j,n)), \tag{4.21}$$

i.e., all but a fraction ε of $I(j,n)$ is within a single ε-ball. Let

$$\hat{I}(j,n) = I(j,n) \cap D.$$

As d is T-invariant, setting $\hat{I}(k,n) = T^{k-j}(\hat{I}(j,n))$, each $\hat{I}(k,n)$ has radius at most ε and occupies a fraction $(1-\varepsilon)$ of $I(k,n)$. As $\varepsilon < 1/10$, and .

$$\mu(T^{N(n)}(I(j,n)) \cap I(j,n)) \geq \frac{1}{3}\mu(I(j,n)),$$

we know

$$\mu(T^{N(n)}(\hat{I}(j,n)) \cap \hat{I}(j,n)) \geq \frac{2}{15}\mu(I(j,n))$$

Hence for a set of points x of positive measure, i.e., those in $\hat{I}(j,n)$,

$$d(T^{N(n)}(x), x) < 2\varepsilon.$$

But as

$$f(x) = d(T^{N(n)}(x), x)$$

is T-invariant it is constant a.e., hence

$$d(T^{N(n)}(x), x) < 2\varepsilon \tag{4.22}$$

almost surely. Using

$$\mu(T^{N(n)+1}(I(j,n)) \cap I(j,n)) \geq \frac{1}{3}\mu(I(j,n))$$

we similarly conclude

$$d(T^{N(n)+1}(x), x) < 2\varepsilon \tag{4.23}$$

almost surely. But then

$$d(T(x), x) < 4\varepsilon \tag{4.24}$$

almost surely, hence $d(T(x), x) = 0$ almost surely and as a.e. orbit is dense for d, d is the trivial pseudometric. ■

Exercise 4.4 Give an alternate proof that Chacón's map is not weakly mixing by showing it has no non-trivial measurable eigenfunctions. Hint: Show that any measurable function must finally be essentially constant on most levels of the towers. Use this, and the 'spacer' to show that the only possible eigenvalue is 1.

Exercise 4.5 Consider the rank-1 cutting and stacking construction with $k(n) = 2^n$, and spacers $S(i,n) = 0$ except for $i = 2^{n-1}$ and $S(2^{n-1}, n) = 1$. Call the ergodic map obtained T.

1. Show T is weakly mixing.
2. Show that there is a sequence $n_k \to \infty$ so that for any set S,

$$\mu(T^{n_k}(S) \cap S) \underset{k}{\rightarrow} \mu(S),$$

(this property is called rigidity).

3. Show that a mixing map is never rigid, and hence T is not mixing.

4. Show that an isometry of a compact metric space is always rigid.

5. Show that Chacón's map is not rigid, and in fact, for no set $S \neq \emptyset$ in X is there a sequence $n_k \nearrow \infty$ with $\mu(T^{n_k}(S) \,\Delta\, S) \to 0$.

   This gives a third proof that Chacón's map is weakly mixing.

4.7 The Kolmogorov property

We end this chapter by giving a definition of K-automorphism (K is for Kolmogorov) that shows how it fits into the hierarchy of mixing conditions. In the next chapter on entropy we will develop a chain of equivalences for the K-automorphisms very similar to that we developed for weak mixing.

Definition 4.5 A map T acting on the Lebesgue probability space $(X, \mathscr{F}, \mu)$ is called a *K-automorphism* provided the following is true.

For any finite partition P and for any $\varepsilon > 0$ there is an N (depending on both P and ε) so that for any $t \in Z^+$ and any integers $n_2, n_3, \ldots, n_t$ with $n_2 > N$, $n_3 - n_2 > N, \ldots, n_t - n_{t-1} > N$, we have

$$\left\| E\left(B \,\middle|\, \bigvee_{i=2}^{t} T^{n_i}(P)\right) - \mu(B) \right\|_1 < \varepsilon \tag{4.25}$$

for all $B \in P$.

Comments on the definition.

1. Notice that if t were kept bounded by k the definition reduces to

$$\left| \frac{\mu(B \cap T^{n_2}(B_2) \cap T^{n_3}(B_3) \ldots \cap T^{n_k}(B_k)}{\mu(T^{n_2}(B_2 \cap T^{n_3}(B_3) \ldots \cap T^{n_k}(B_k))} - \mu(B) \right| < \varepsilon \tag{4.26}$$

once $n_2, n_3 - n_2, \ldots, n_k - n_{k-1}$ are large enough. A little induction shows that this is equivalent to k-fold mixing. Thus the K-property is a kind of uniform k-fold mixing, and K-automorphisms are mixing of all orders.

2. Although the definition gives uniformity in t, we must choose the sets B, $B_2, \ldots, B_t$ from some fixed finite collection.

   If we allowed ourselves to choose from an infinite collection, for example, B, $T^{-1}(B)$, $T^{-2}(B)$, $\ldots$, no map would satisfy the condition.

3. We use conditional expectation in the definition as opposed to

$$\mu(B \cap T^{n_2}(B_2) \cap \cdots \cap T^{n_t}(B_t))$$

as these values automatically tend to zero in t if T is mixing. For any ε and for large t, the condition would be vacuous. Thus this property is just k-fold mixing for all k.

The K-property asks for more, requiring that each new $T^{n_t}(B_t)$ mix well with the previous ones relative to their size. Just as weakly mixing opened the door to spectral theory and point spectrum, the K-property will open the door to entropy.

5 Entropy

Our approach so far to the classification of ergodic processes has been to search for natural invariants of the process. We have discussed two sorts, the point spectrum and mixing properties. One more invariant remains to be looked at, perhaps the most fundamental numerical invariant of stationary stochastic processes, the Kolmogorov–Sinai entropy. What the entropy attempts to measure is the rate at which a process becomes random.

5.1 Counting names

The approach we take to this question is unusual, being based from the outset on name counting. We will not intersect the classical theory for some time (at Definition 5.3). With our approach the ideas will evolve quite naturally, and the skills we develop in manipulating names will stand us in good stead in Chapters 6 and 7.

Let's begin to focus precisely on how entropy is computed. Let T, acting on the Lebesgue probability space $(X, \mathscr{F}, \mu)$ be an invertible measure-preserving ergodic map, and $P = \{p_1, \ldots, p_s\}$ be a finite partition of X into measurable sets.

Here is a slightly novel notion of a 'partition' which will be very useful for us. Regard P as a function from X to a 'finite state space' of 'symbols' or 'names.' Thus $P: X \to \{p_1, \ldots, p_s\}$ is a finite partition and $\{p_1, \ldots, p_s\}$ is the space of symbols. The set p_k is then more precisely $P^{-1}(p_k)$, the set of points 'named,' or 'labeled' p_k. It will happen often that we will label several partitions with the same labels. In this case the state space will be lower case letters, subscripted to order them. The function will be indicated by the corresponding capital letter, embellished with a subscript or other notational device to indicate it uniquely ($P_1, P_2, P', \bar{P}$, etc.). When there is no ambiguity we will refer to a set by its name.

The *T,P,n-name* of $x \in X$ is a sequence of n symbols, each chosen from among $p_1, \ldots, p_s$ written

$$\mathbf{p}_n(x) = (P(x), P(T(x)), \ldots, P(T^{n-1}(x))) \tag{5.1}$$

where $P(x)$ is that element of P containing x.

Thus $\mathbf{p}_n$ maps X to P^n. These sequences of n symbols from P are then the names for the sets in $\bigvee_{i=0}^{n-1} T^{-i}(P)$. The measure μ on X is transported by $\mathbf{p}_n$ to a measure on the elements $\mathbf{P}^n$—the measure of a name is the measure of the set of points which have that name.

Entropy, roughly speaking, is the exponential rate of growth in n of the number of T,P,n-names. To be precise, fix $\varepsilon > 0$. Starting from the names of least measure in P^n, remove as many as possible so that the measure of the remaining names is still greater than $(1 - \varepsilon)$. This collection of remaining names we denote by

$$S(T, P, n, \varepsilon).$$

The presence of the ε is what ties the entropy to the invariant measure μ. We only consider the 'large' names. If we did not omit any names, measure would play no role in the definition. This is a fruitful approach also and leads to the development of topological entropy for symbolic systems.

Let

$$N(T, P, n, \varepsilon) = \#(S(T, P, n, \varepsilon)). \tag{5.2}$$

It seems reasonable to guess that this number would grow exponentially in n. To try to extract the coefficient of this growth we introduce the following definition.

Definition 5.1 For T ergodic and P a finite partition let

$$h(T, P, n, \varepsilon) = \frac{1}{n}\log_2(N(T, P, n, \varepsilon)) \quad [\text{note: } \leq \log_2 s]. \tag{5.3}$$

To remove dependence on n set

$$h(T, P, \varepsilon) = \varliminf_{n\to\infty} h(T, P, \varepsilon, n), \tag{5.4}$$

and to remove dependence on ε set

$$h(T, P) = \lim_{\varepsilon\to 0} h(T, P, \varepsilon). \tag{5.5}$$

The limit certainly exists as the sequence $h(T, P, \varepsilon)$ is monotone increasing (as $\varepsilon \to 0$) and bounded (by $\log_2 s$).

We first use the backward Vitali lemma to show that ε plays no significant role, except to be less than 1, in this computation.

Theorem 5.1 *For T ergodic and P a finite partition, for any ε', with $\varepsilon' < \varepsilon < 1$ we have*

$$\varlimsup_{n\to\infty} h(T, P, n, \varepsilon') \leq \varliminf_{n\to\infty} h(T, P, n, \varepsilon). \tag{5.6}$$

Proof Fix ε, ε' and select $\{n_i\}$ so that

$$\lim_{i\to\infty} h(T, P, n_i, \varepsilon) = \varliminf_{n\to\infty} h(T, P, n, \varepsilon) = h(T, P, \varepsilon).$$

Also fix $\bar{\varepsilon} < \varepsilon'$, and assume $|h(T, P, n_i, \varepsilon) - h(T, P, \varepsilon)| < \bar{\varepsilon}/10$.

Recall that $S(T, P, n_i, \varepsilon)$ is the set of 'large' names—i.e., a minimal (in cardinality) set of T,P,n_i-names sufficient to cover all but ε (in measure) of X.

Set

$$\bar{S}_i = \{x : \mathbf{p}_{n_i}(x) \in S(T, P, n_i, \varepsilon)\}. \tag{5.7}$$

Clearly $\mu(\bar{S}_i) \geq 1 - \varepsilon$.

Let

$$\bar{S} = \overline{\lim_i}\, \bar{S}_i$$

[$= \bigcap_{N=1}^{\infty} \bigcup_{i=N}^{\infty} \bar{S}_i =$ the subset of those x which lie in infinitely many $\bar{S}_i$], and $\mu(\bar{S}) \geq 1 - \varepsilon > 0$.

Consider

$$\hat{S} = \bigcup_{i=0}^{\infty} T^i(\bar{S}).$$

Since T is ergodic $\mu(\hat{S}) = 1$ and without loss of generality we may assume $\hat{S} = X$.

For $x \in X = \hat{S}$ we have $x \in T^i(\bar{S})$ for some i so $T^{-i}(x) \in \bar{S}$ and for infinitely many k, $T^{-i}(x) \in \bar{S}_k$. Let $k(1, x) < k(2, x) < \cdots$ be these k's.

Once $n_{k(n,x)} > i$ we may set $i_n = -i$, $j_n = n_{k(n,x)} - i$. Thus for each $x \in X$ there are values $i_n < 0$, $j_n > 0$ with $j_n - i_n \to \infty$ in n and the $T, P, j_n - i_n$-name of $T^{-i_n}(x)$ is in the set $S(T, P, j_n - i_n, \varepsilon)$. We also know $|h(T, P, j_n - i_n, \varepsilon) - h(T, P, \varepsilon)| < \bar{\varepsilon}/10$.

We are now ready to use the backward Vitali lemma (Theorem 3.9). Applying it, there is a set F, and for $x \in F$, values $i(x)$, $j(x)$ from among the $i_n(x)$, $j_n(x)$ so that the orbit intervals

$$\{T^{i(x)}(x), T^{i(x)+1}(x), \ldots, T^{j(x)}(x)\}$$

are disjoint and cover all but $\bar{\varepsilon}/10$ of X. There is also an N_0 with $j(x) - i(x) < N_0$ uniformly.

Let

$$G = \bigcup_{x \in F} \{T^{i(x)}(x), \ldots, T^{j(x)}(x)\}$$

and $\mu(G) > 1 - \bar{\varepsilon}/10$.

Select $N \geq 10N_0/\bar{\varepsilon}$ and so large that for $n \geq N$, for all but $\bar{\varepsilon}$ of X in measure and for all but a set of density at most $\bar{\varepsilon}/5$ of $i \in \{0, 1, \ldots, n-1\}$, we have $T^i(x) \in G$. That we can do this is a consequence of the Birkhoff ergodic theorem applied to χ_G. Let H be this good set for the ergodic theorem.

We now compute an upper bound for the number of T,P,n-names for points in H. Such a T,P,n-name can be represented as in Fig. 5.1. There is a subset of the name of density at least $1 - (2/5)\bar{\varepsilon}$ consisting of disjoint blocks $I_1, I_2, \ldots, I_l$ and across each such block we see a name from $S(T, P, \# I_k, \varepsilon)$. These blocks are the intervals

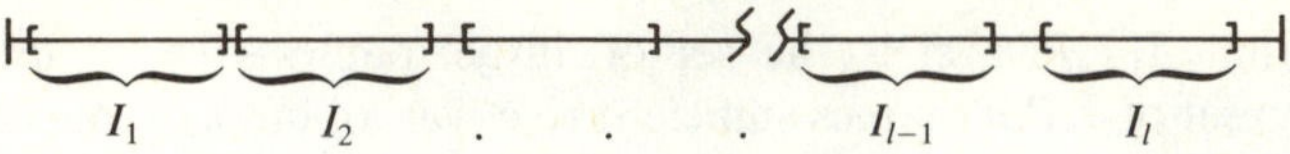

Fig. 5.1 T,P,n-names for x.

$$(i(T^u(x)) + u, j(T^u(x) + u)$$

where $T^u(x) \in F$, and the entire block is contained in $(0, 1, \ldots, n-1)$.

The remaining indices correspond to points $T^u(x) \notin G$ or whose block of indices is not completely contained in $(0, \ldots, n-1)$.

The number of such T,P,n-names in H is bounded by

$$\begin{pmatrix} \#\text{ of ways} \\ \text{the } I_1, \ldots, I_l \\ \text{can arise} \end{pmatrix} \times \prod_{k=1}^{l} \begin{pmatrix} \#\text{ of names} \\ \text{possible} \\ \text{across } I_k \end{pmatrix} \times \begin{pmatrix} \#\text{ of names} \\ \text{possible out-} \\ \text{side the } I_k \end{pmatrix}. \tag{5.8}$$

Now

$$(1) \quad \begin{pmatrix} \#\text{ of ways} \\ \text{the } I_1, \ldots, I_l \\ \text{can arise} \end{pmatrix} \leq \begin{pmatrix} \#\text{ of subsets of size} \\ \text{at most } \tfrac{2}{5}\bar{\varepsilon}n \text{ in a} \\ \text{set of size } n \end{pmatrix},$$

$$(2) \quad \begin{pmatrix} \#\text{ of names} \\ \text{possible} \\ \text{across } I_k \end{pmatrix} = 2^{h(T, P, \# I_k, \varepsilon)(\# I_k)}$$

and

$$(3) \quad \begin{pmatrix} \#\text{ of names} \\ \text{possible out-} \\ \text{side the } I_k \end{pmatrix} \leq s^{2/5\bar{\varepsilon}n}. \tag{5.9}$$

To estimate the number of subsets of size at most $(2/5)\bar{\varepsilon}n$ in a set of size n we use Stirling's formula.

$$n! = \left(\frac{\mathrm{n}}{\mathrm{e}}\right)^n \mathrm{e}^{\zeta(n)} \sqrt{2\pi/n} \tag{5.10}$$

where

$$0 < \zeta(n) < \frac{1}{12n}.$$

Note that this fundamental combinatorial formula can be regarded as the core of entropy theory. See Ahlfors (1966) for a proof.

Thus the binomial coefficient

$$\binom{n}{k} < \left(\left(\frac{k}{n}\right)^{-k/n}\left(\frac{(n-k)}{n}\right)^{-(n-k)/n}\right)^n \sqrt{\frac{k(n-k)}{n}}\,\mathrm{e}^{\zeta(n)} \tag{5.11}$$

So the number of sets of size at most αn in a set of size n is

$$\binom{n}{[\alpha n]} + \binom{n}{[\alpha n]-1} + \cdots + \binom{n}{1} + \binom{n}{0} \le \left(\alpha^{-\alpha}(1-\alpha)^{-(1-\alpha)}\right)^n \sqrt{n}. \tag{5.12}$$

Set

$$H(\alpha) = -\alpha \log_2 \alpha - (1-\alpha)\log_2(1-\alpha). \tag{5.13}$$

Combining estimates (5.11) and (5.12), the number of T,P,n-names covering all but $\bar{\varepsilon}$ of X is bounded by 2^r, where

$$r = H\left(\frac{2\bar{\varepsilon}}{5}\right)(n + c\log_2 n) + \left(\frac{2\bar{\varepsilon}}{5}\right)(\log_2 s)n + \left(h(T,P,\varepsilon) + \frac{\bar{\varepsilon}}{10}\right)\sum_{k=1}^{\ell} \# I_k$$

$$\le \left(h(T,P,\varepsilon) + \frac{\bar{\varepsilon}}{10} + H\left(\frac{2\bar{\varepsilon}}{5}\right) + c\frac{\log_2 n}{n} + \left(\frac{2\bar{\varepsilon}}{5}\right)\log_2 s\right)n.$$

Thus

$$h(T,P,n,\bar{\varepsilon}) \le h(T,P,\varepsilon) + \frac{c\log_2 n}{n} + \frac{\bar{\varepsilon}}{10} + H\left(\frac{2\bar{\varepsilon}}{5}\right) + \frac{2\bar{\varepsilon}}{5}\log_2 s.$$

This holds for all n sufficiently large so

$$\overline{\lim_{n\to\infty}}\, h(T,P,n,\bar{\varepsilon}) \le h(T,P,\varepsilon) + \frac{\bar{\varepsilon}}{10} + H\left(\frac{2\bar{\varepsilon}}{5}\right) + \frac{2\bar{\varepsilon}}{5}\log_2 s. \tag{5.14}$$

Clearly for $\bar{\varepsilon} \le \varepsilon'$,

$$\overline{\lim_{n\to\infty}}\, h(T,P,n,\bar{\varepsilon}) \ge \overline{\lim_{n\to\infty}}\, h(T,P,n,\varepsilon')$$

so if we let $\bar{\varepsilon} \to 0$ in (5.14),

$$\overline{\lim_{n\to\infty}}\, h(T,P,n,\varepsilon') \le h(T,P,\varepsilon).$$

It follows now that for $\varepsilon < 1$,

$$\lim_{n\to\infty} h(T,P,n,\varepsilon)$$

exists and its value is $h(T,P)$ independent of ε.

Corollary 5.2 (Of the proof) *Given T as usual and a finite partition P, there exists an increasing sequence of sets A_n whose limit is a.a. of X so that*

$$\frac{\log_2(\#\text{ of } T,P,n\text{-names in } A_n)}{n} \tag{5.15}$$

converges to $h(T,P)$ as $n \to \infty$.

Proof In the course of proving the previous theorem we selected $\bar{\varepsilon}$ and constructed a set G of measure at least $1 - \bar{\varepsilon}/10$ using the backward Vitali lemma. We discarded those points of X whose orbit of length n was outside G more than $\bar{\varepsilon}/5$ of the time—by the pointwise ergodic theorem this is a decreasing sequence of sets. More precisely, let

$$B_{n,\varepsilon} = \left\{ x : \text{pointwise ergodic theorem holds for } \chi_G \right.$$
$$\left. \text{for all } n' \geq n \text{ to within an error of } \frac{\bar{\varepsilon}}{10} \right\}.$$

Here we choose $\bar{\varepsilon}$ so

$$H\left(\frac{2\bar{\varepsilon}}{5}\right) + \frac{2\bar{\varepsilon}}{5}\log_2(s) + \frac{\bar{\varepsilon}}{10} < \frac{\varepsilon}{2}$$

and n so large that

$$\frac{c \log_2 n}{n} < \frac{\varepsilon}{2}.$$

By the previous proof,

$$\frac{\log(\# T,P,n'\text{-names in } B_{n,\varepsilon})}{n'} = h(T, P) \pm \varepsilon$$

for all $n' \geq n$.

Now let $\varepsilon_i = 2^{-i}$ and select $\{n_i\}$ increasing so that

$$\mu(B_{n_i,\varepsilon_i}) > 1 - 2^{-i}$$

and for $n' > n_i$

$$\log\left(\frac{(\# T,P,n'\text{-names in } B_{n_i,\varepsilon_i})}{n'}\right) = h(T, P) \pm \varepsilon_i.$$

Set

$$A_n = \bigcap_{n_{i+1} \geq n} B_{n_i,\varepsilon_i}.$$

Now

$$\mu(A_n) > 1 - 2^{-i},$$

so A_n increases to a.a. of X. Also

$$\frac{\log(\# T,P,n\text{-names in } A_n)}{n} \leq \frac{\log(\# T,P,n\text{-names in } B_{n_i,\varepsilon_i})}{n}$$
$$= h(T, P) \pm \varepsilon_i$$

where

$$n_i < n \le n_{i+1}.$$

Hence

$$\overline{\lim} \frac{\log(\# T,P,n\text{-names in } A_n)}{n} \le h(T,P).$$

However,

$$\underline{\lim} \frac{\log(\# T,P,n\text{-names in } A_n)}{n} \ge h(T,P)$$

anyway as $\mu(A_n)\,\mathrm{d}\mu \overset{n}{\to} 1$. ■

5.2 The Shannon–McMillan–Breiman theorem

We have often spoken of the measure of a T,P,n-name, meaning the measure of the set of points with this T,P,n-name. We write this as $\mu(\mathbf{p}_n(x))$ allowing $\mathbf{p}_n(x)$ to represent both the name and the set of points possessing it. Our next result concerns the asymptotic size of such names.

Theorem 5.3 (Shannon–McMillan–Breiman) *For T an ergodic map and P a finite partition, for a.e. $x \in X$*

$$\lim_{n\to\infty} \frac{-\log_2(\mu(\mathbf{p}_n(x)))}{n} = h(T,P). \tag{5.16}$$

Note: this convergence is also in $L^1(\mu)$.

Proof We first show

$$\overline{\lim_{n\to\infty}} \frac{-\log_2(\mu(\mathbf{p}_n(x)))}{n} \le h(T,P) \tag{5.17}$$

for a.e. $x \in X$.

Let

$$B_{n,\varepsilon} = \left\{x : \frac{-\log_2(\mu(\mathbf{p}_n(x)))}{n} > h(T,P) + \varepsilon, x \in A_n\right\}$$

where A_n is the set constructed in the previous corollary.

Once n is sufficiently large

$$\mu(B_{n,\varepsilon}) \le 2^{-(h(T,P)+\varepsilon)n} \cdot 2^{(h(T,P)+(\varepsilon/2))n} = 2^{-\varepsilon n/2}.$$

Thus

$$\sum_{n=1}^{\infty} \mu(B_{n,\varepsilon}) < \infty.$$

By the Borel–Cantelli lemma

$$\mu\{x : x \text{ lies in infinitely many } B_{n,\varepsilon}\} = 0.$$

If x lies in only finitely many $B_{n,\varepsilon'}$ then for large n either $x \notin A_n$ or

$$\frac{-\log_2(\mu(\mathbf{p}_n(x)))}{n} < h(T, P) + \varepsilon.$$

As the A_n increase to a.a. of X, for a.e. x, once n is large enough, $x \in A_n$. Hence for a.e. x, once n is large enough

$$-\frac{1}{n}\log_2 \mu(\mathbf{p}_n(x)) < h(T, P) + \varepsilon.$$

Thus for a.e. x,

$$\overline{\lim_{n\to\infty}}\left(-\frac{1}{n}\log_2 \mu(\mathbf{p}_n(x))\right) < h(T, P),$$

which is (5.17). Now we prove the lower estimate

$$\underline{\lim_{n\to\infty}} -\frac{1}{n}\log_2 \mu(\mathbf{p}_n(x)) \geq h(T, P). \tag{5.18}$$

Redefine

$$B_{n,\varepsilon} = \left\{x : -\frac{1}{n}\log \mu(\mathbf{p}_n(x)) < h(T, P) - \varepsilon\right\}.$$

Thus the number of T,P,n-names in $B_{n,\varepsilon}$ is less than or equal to $2^{(h(T,P)-\varepsilon)n}$. Let

$$B_\varepsilon = \{x : x \text{ is an infinitely many } B_{n,\varepsilon}\}.$$

If $\mu(B_\varepsilon) = 0$ for all ε we are done.

Assume $\mu(B_\varepsilon) > 0$ for some $\varepsilon > 0$. Now by ergodicity

$$\mu \bigcup_{i=0}^{\infty} T^i(B_\varepsilon) = 1,$$

and for a.e. x there exist $i_n(x) < 0$, $j_n(x) > 0$ with $j_n(x) - i_n(x) \xrightarrow{n} \infty$ and such that the $T,P,j_n(x) - i_n(x)$-name of $T^{i_n(x)}(x)$ is among the $2^{(h(T,P)-\varepsilon)(j_n(x)-i_n(x))}$ names in $B_{j_n(x)-i_n(x),\varepsilon}$.

Following the sequence of estimates of Theorem 5.1, using the backward Vitali lemma we can conclude from this that $h(T, P) < h(T, P) - \varepsilon/2$, a contradiction. Hence $\mu(B_\varepsilon) = 0$ for all $\varepsilon > 0$ and (5.18) holds finishing the proof of pointwise convergence. To see uniform integrability and hence $L^1(\mu)$ convergence, just notice

$$\mu\left\{x : -\frac{1}{n}\log_2 \mu(\mathbf{p}_n(x)) > \log_2 s + a\right\} \leq 2^{-na}. \qquad \blacksquare$$

5.3 Entropy zero and past algebras

To this point entropy has been tied to a fixed finite partition P. We need to loosen this tie. Our first step is to understand what $h(T, P) = 0$ means. This in itself is an important step. Later we shall see that the relationship between entropy zero and the K-property is very analogous to the relationship between isometries and the weakly mixing property.

Let T be, as usual, an invertible measurable measure-preserving map on the Lebesgue probability space $(X, \mathscr{F}, \mu)$, and $P = \{p_1, \ldots, p_s\}$ a finite partition of X.

Define the *past* σ-algebra of T, P to be

$$\mathscr{P} = \bigvee_{j=-1}^{-\infty} T^{-j}(P), \tag{5.19}$$

and more generally

$$\mathscr{P}_{u,v} = \bigvee_{j=u}^{v} T^{-j}(P), \tag{5.20}$$

where $u \leq v$. If $(u', v') \subseteq (u, v)$ then $\mathscr{P}_{u',v'} \subseteq \mathscr{P}_{u,v}$. The sets in $\mathscr{P}_{u,v}$ consist of points whose T,P-names agree from index u to index v. This T,P-name we write $\mathbf{p}_{u,v}(x)$.

Let A be any measurable set. We want to define *the conditional expectation of A given $\mathscr{P}$* which we will write $E(A|\mathscr{P})$.

Define

$$f_N(x) = \frac{\mu(A \cap \mathbf{p}_{-1,-N}(x))}{\mu(\mathbf{p}_{-1,-N}(x))} = E(A|P_{-1,-N}).$$

It is an easy check that the functions f_N and algebras $P_{-1,-N}$ form a bounded positive martingale. The L^1-martingale theorem (Corollary 2.8) of Chapter 2 tells us, for a.e. x, $f_N(x)$ converges. We call the limit functions, defined a.e.,

$$E(A|\mathscr{P}). \tag{5.21}$$

We know, for any $S \in \mathscr{P}$,

$$\int_S \chi_A \,\mathrm{d}\mu = \int_S E(A|\mathscr{P}) \,\mathrm{d}\mu. \tag{5.22}$$

Suppose $A = p_i$, an element of the partition P. The function $E(p_i|\mathscr{P})$ measures the probability that, given only the past history of a point x, that it now lies in p_i.

Note that since $\sum_{p_i \in p} \chi_{p_i} = 1$, we have

$$\sum_{p_i \in p} E(p_i|\mathscr{P}) = 1 \quad \text{a.s.},$$

so for a.e. x, the vector

$$(E(p_1|\mathscr{P})(x), E(p_2|\mathscr{P})(x), \ldots, E(p_s|\mathscr{P})(x))$$

forms a probability vector which we write

$$D(P|\mathscr{P})(x), \tag{5.23}$$

a probability distribution-valued function.

Theorem 5.4 *$h(T, P) = 0$ iff for all $p_i \in P$, $E(p_i|\mathscr{P}) = \chi_{p_i}$ a.s. i.e., the past of the (T, P) process determines its present.*

Proof Suppose that for all $p_i \in P$,

$$E(p_i|\mathscr{P}) = \chi_{p_i} \quad \text{a.s.}$$

For $x \in X$, let $p(x) \in P$ be that element which contains x. Thus for a.e. x, $E(p(x)|\mathscr{P}) = 1$.

Fix $\varepsilon > 0$ and select N_0 so large that on a set G, $\mu(G) > 1 - \varepsilon$, we have for $x \in G$

$$G \cap \mathbf{P}_{-1,-N_0}(x) \subset p(x).$$

Exercise 5.1 How do we do this?

Thus if we know that $x \in G$ and we know the name $\mathbf{P}_{-1,-N_0}(x)$ then we know $p(x)$. By the ergodic theorem we can select N so large that for all but ε of the points $x \in X$, at least $(1 - 2\varepsilon)N$ of $x, T(x), \ldots, T^{N-1}(x)$ are in G. Call this subset B_N,

$$\mu(B_N) > 1 - \varepsilon.$$

We now count the number of T,P,N-names in H_N. We do this by first selecting a subset of $2\varepsilon N$ places in $0, \ldots, N - 1$ which are to be those indices not in G. At these indices in the name, and in the first N_0 positions we assign some arbitrary symbols from P. The rest of the symbols in the name are now determined, as working inductively from the left, at an undetermined index we must be in G and we know symbols in the previous N_0 positions.

Thus

$$\#(T,P,N\text{-names in } B_N)$$
$$\le \begin{pmatrix} \#\text{ of subsets of} \\ \le 2\varepsilon N \text{ in a set} \\ \text{of } N \end{pmatrix} \times \begin{pmatrix} \#\text{ names} \\ \text{across} \\ \text{such a set} \end{pmatrix} \times \begin{pmatrix} \#\text{ names} \\ \text{across} \\ (0, \ldots, N_0 - 1) \end{pmatrix} \tag{5.24}$$
$$\le 2^{(H(2\varepsilon) + c \log N/N + 2\varepsilon \log s + (N_0/N) \log s)N}$$

(cf. estimates (5.11), (5.12) and (5.14)). Thus

$$h(T, P, \varepsilon, N) \le H(2\varepsilon) + \frac{c \log N}{N} + 2\varepsilon \log s + \frac{N_0}{N} \log s.$$

Letting $N \to \infty$ and then $\varepsilon \to 0$,

$$h(T, P) = 0.$$

To prove the converse we will show that if $E(p_i|\mathscr{P})$ is not 0, 1 a.s., then $h(T, P) > 0$.

Note: To say

$$E(p_i|\mathscr{P}) = \begin{cases} 0 \\ 1 \end{cases} \quad \text{a.s.}$$

is the same as to say

$$D(p_i|\mathscr{P}) = \chi_{p_i} \quad \text{a.s.} \tag{5.25}$$

Since there are only finitely many elements p_i in P, for some fixed p_i and for some $\alpha > 0$ we must have $1 - \alpha > E(p_i|\mathscr{P}) > \alpha$ on a set $A \in \mathscr{P}$, $\mu(A) > \alpha$.

Select $\varepsilon < \alpha$ and N_0 so large that for a set G of measure $\mu(G) > 1 - \varepsilon$ we have for all $x \in G$ and $n \geq N_0$,

$$E(p_i|\mathscr{P})(x) - E(p_i|P_{-1,-N})(x) < \varepsilon. \tag{5.26}$$

This follows from the pointwise convergence of the martingale

$$f_n = E(p_i|P_{-1,-N}).$$

This now gives us information about the asymptotic behavior of $\mu(\mathbf{p}_n(x))$. The key observation is that if $T^n(x) \in A \cap G$ and $n \geq N_0$, then

$$\frac{\mu(\mathbf{p}_{n+1}(x))}{\mu(\mathbf{p}_n(x))} < 1 - \alpha + \varepsilon. \tag{5.27}$$

To see this just note

$$\frac{\mu(\mathbf{p}_{n+1}(x))}{\mu(\mathbf{p}_n(x))} = E(P(T^n(x))|\mathbf{p}_{-1,-n}(T^n(x)))$$

and as $T^n(x) \in G \cap A$,

$$\begin{aligned} E(P(T^n(x))|\mathbf{p}_{-1,-n}(T^n(x))) &= E(P(T^n(x))|\mathscr{P}) \pm \varepsilon \\ &< \begin{cases} E(p_i|\mathscr{P}) + \varepsilon & \text{if } P(T^n(x)) = p_i \\ 1 - E(p_i|\mathscr{P}) + \varepsilon & \text{if } P(T^n(x)) \neq p_i \end{cases} \\ &< 1 - \alpha + \varepsilon. \end{aligned}$$

Regardless of whether $T^n(x)$ is in $A \cap G$ or not,

$$\frac{\mu(\mathbf{p}_{n+1}(x))}{\mu(\mathbf{p}_n(x))} \leq 1.$$

Write

$$\mu(\mathbf{p}_n(x)) = \mu(P(x)) \cdot \sum_{i=1}^{N-1} \frac{\mu(\mathbf{p}_{i+1}(x))}{\mu(\mathbf{p}_i(x))}. \tag{5.28}$$

Each factor is less than or equal to 1 and whenever $i \geq N_0$ and $T^i(x) \in A \cap G$ it is less than or equal to $1 - \alpha + \varepsilon$.

Setting

$$k(n, x) = \sum_{i-N_0}^{N} \chi_{A\cap G}(T^i(x)),$$

we have

$$\mu(\mathbf{p}_n(x)) \leq (1 - \alpha + \varepsilon)^{k(N,x)}.$$

By the Birkhoff theorem, for a.e. x

$$\lim_{N\to\infty} \frac{1}{N} k(n, x) = \mu(G \cap A) > \alpha - \varepsilon$$

and so for a.e. x

$$\overline{\lim_{N\to\infty}} \frac{-\log_2(\mu(\mathbf{p}_n(x)))}{n} \geq (\alpha - \varepsilon)\log_2(1 - \alpha + \varepsilon) > 0.$$

But we already know this limit exists (Theorem 5.3) and is $h(T, P)$. ■

5.4 More about the *K*-property

We now turn to the opposite extreme, that of transformations T for which $h(T, P) > 0$ for all non-trivial partitions P. We will see that these are the K-systems of the last chapter.

Theorem 5.5 *If T is a K-system then $h(T, P) > 0$ for all finite non-trivial partitions P.*

This theorem depends on two lemmas.

Lemma 5.6 *If $h(T, P) = 0$, then for all integers n, $h(T^n, P) = 0$.*

Proof Certainly $N(T^n, P, k, \varepsilon) \leq N(T, P, nk, \varepsilon)$ and $h(T^n, P) \leq nh(T, P)$. ■

Lemma 5.7 *If T is a K-system, P a non-trivial partition and $\varepsilon > 0$, then there is an n so that for a set G,*

$$\mu(G) > 1 - \varepsilon.$$

If $x \in G$ then

$$E\left(p_i \middle| \bigvee_{k=-1}^{-\infty} T^{-nk}(P)\right) = \mu(p_i) \pm \varepsilon$$

for all $p_i \in P$.

Proof We know

$$E\left(p_i \,\middle|\, \bigvee_{k=-1}^{-\infty} T^{-nk}(P)\right) = \lim_{N\to\infty} E\left(p_i \,\middle|\, \bigvee_{k=-1}^{-N} T^{-nk}(P)\right)$$

pointwise a.e. For any $x \in X$,

$$E\left(p_i \,\middle|\, \bigvee_{k=-1}^{-N} T^{-nk}(P)\right)(x) = \frac{\mu(p_i \cap P(T^{-n}(x)) \cap P(T^{-2n}(x)) \cap \cdots \cap P(T^{-Nn}(x)))}{\mu(P(T^{-n}(x)) \cap P(T^{-2n}(x)) \cap \cdots \cap P(T^{-Nn}(x)))}.$$

As T is a K-system, (see Definition 4.4) once n is sufficiently large, for all N, for all but ε of X, this is $\mu(p_i) \pm \varepsilon$ for all p_i. ■

Proof of Theorem 5.5 If P is a non-trivial partition, as T is a K-system by Lemma 5.7 there is an n so that $E(p_i | \bigvee_{k=-1}^{-\infty} T^{-nk}(P)) \neq 0$ or 1 with positive probability (in fact this conditional expectation is approximately $\mu(p_i(x))$ for most x). This says $h(T^n, P) \neq 0$. Now by our first lemma $h(T, P) \neq 0$.

5.5 The entropy of an ergodic transformation

We want to prove the converse of Theorem 5.5 but it will be some time before we have sufficient machinery to do so. We now begin to develop this machinery, for a time leaving behind the notion of K-system and re-entering the general development of the theory of entropy.

Definition 5.2

$$h(T) = \sup h(T, P) \tag{5.29}$$

where the sup is taken over all finite partitions P of X.

We need some basic facts to help make $h(T)$ a reasonably computable quantity.

Lemma 5.8 *For all $k > 0$,*

$$h\left(T, \bigvee_{i=-k}^{k} T^{-i}(P)\right) = h(T, P). \tag{5.30}$$

Proof We estimate the number of $T, \bigvee_{i=-k}^{k} T^{-i}(P)$, n-names on a subset A. On the one hand it is at least the number of T,P,n-names on A. On the other hand to know the $T, \bigvee_{i=-k}^{-k} T^{-i}(P)$, n-name of x is precisely to know $\mathbf{p}_{-k,n+k}(x)$, hence there are at most s^{2k} times as many $T, \bigvee_{i=-k}^{k} T^{-i}(P)$, n-names on A as there are T,P,n-names.

Hence

$$\#(T,P,n\text{-names in } A) \le \#\left(T, \bigvee_{i=-k}^{k} T^{-i}(P), n\text{-names in } A\right)$$
$$\le s^{2k}\,\#(T,P,n\text{-names in } A).$$

The result follows. ■

Lemma 5.9 *Suppose a partition H is P measurable, i.e., an element of H is a union of elements of P. Then*

$$h(T,H) \le h(T,P). \tag{5.31}$$

Proof Exercise 5.2. ■

Exercise 5.2 Prove Lemma 5.9.

If we have two finite partitions of X, H, and H', both with state space $\{h_1,\ldots,h_s\}$, we can define their symmetric differences $H \,\Delta\, H' = \{x : H(x) \ne H'(x)\}$. This set is a measure of how different the partitions are. Notice it *does* take into account the labels on the sets.

Lemma 5.10 *Suppose H and H' are two finite partitions with the same state space $\{h_1,\ldots,h_s\}$. If $\mu(H \,\Delta\, H') < \varepsilon$, i.e., H and H' are very close, then*

$$h(T,H) \le h(T,H') + \varepsilon \log_2 s + H(\varepsilon). \tag{5.32}$$

Proof Fix $\bar{\varepsilon} > 0$ and select N_1 so large that for a set A_1, $\mu(A_1) > 1 - \bar{\varepsilon}$ and for $n \ge N_1$,

$$\#(T,H',n\text{-names in } A_1) \le 2^{(h(T,H)+\bar{\varepsilon})n}$$

(cf. Theorem 5.3).

Let $E = \{x : H(x) \ne H'(x)\}$, and hence $\mu(E) < \varepsilon$. Applying the Birkhoff theorem to E, select N_2 so large that for a set A_2 with $\mu(A_2) > 1 - \bar{\varepsilon}$ we have for $x \in A_2$ and $n \ge N_2$, for all but at most $(\varepsilon + \bar{\varepsilon})n$ of the points x, $T(x)$, …, $T^{n-1}(x)$, $H(T^j(x)) = H'(T^j(x))$.

Let $A = A_1 \cap A_2$, so $\mu(A) \ge 1 - 2\bar{\varepsilon}$; let $N = \max\{N_1, N_2\}$, and $n \ge N$. Then

$$2^{h(T,H,2\bar{\varepsilon},n)} \le \#(T,H,n\text{-names in } A)$$
$$\le \#\; T,H',n\text{-names in } A)$$
$$\times \begin{pmatrix} \#\text{ of subsets of size} \le (\varepsilon + \bar{\varepsilon})n \\ \text{in a set of size } n \end{pmatrix} \times s^n.$$

Applying the usual Stirling's formula estimates

$$h(T,H,2\bar{\varepsilon},n) \le h(T,H') + \bar{\varepsilon} + H(\varepsilon + \bar{\varepsilon}) + \frac{c \log n}{n} + (\varepsilon + \bar{\varepsilon}) \log s.$$

Letting $n \to \infty$ and $\bar{\varepsilon} \to 0$ we get the conclusion. ■

Corollary 5.11 *If*

$$H \subset \bigvee_{i=-\infty}^{\infty} T^{-i}(P)$$

then

$$h(T, H) \leq h(T, P).$$

Proof For any $\varepsilon > 0$ select k and

$$H' \subset \bigvee_{i=-k}^{k} T^{-i}(P)$$

with

$$\mu(H \,\Delta\, H') < \varepsilon.$$

Now applying Lemmas 5.9 and 5.10

$$\begin{aligned} h(T, H) &\leq h(T, H') + H(\varepsilon) + \varepsilon \log s \\ &\leq h\left(T, \bigvee_{i=-k}^{k} T^{-i}(P)\right) + H(\varepsilon) + \varepsilon \log s \\ &\leq h(T, P) + H(\varepsilon) + \varepsilon \log s. \end{aligned}$$

Letting $\varepsilon \to 0$ completes the result. ∎

Corollary 5.12 *If P is a generating partition, i.e.,*

$$\bigvee_{i=-\infty}^{\infty} T^{-i}(P) = \mathscr{F}$$

then

$$h(T) = h(T, P).$$ ∎

5.6 Examples of entropy computations

Example 1 *Irrational Rotations.* Let $T = R_\alpha$, an irrational rotation of the circle

$$\theta \to (\theta + \alpha) \bmod 2\pi.$$

The partition

$$P = \{(0, \pi), [\pi, 2\pi]\}$$

is a generator. In fact, if $\theta_1 \neq \theta_2$, as the values $n\alpha \bmod 2\pi$ are dense in $[0, 2\pi)$, for some n, $(n\alpha) \bmod 2\pi$ lies between θ_1 and θ_2, hence θ_1 and θ_2 lie in distinct elements of $T^n(P)$. Thus $\bigvee_{i=-\infty}^{\infty} T^{-i}(P)$ separates points and so generates. Thus $h(R_\alpha) = h(R_\alpha, P)$.

Theorem 5.13 $h(R_\alpha, P) = 0$.

Proof A set in

$$\bigvee_{i=0}^{n} T^{-i}(P)$$

is an interval. Spanning this with $T^{-n}(P)$ cuts exactly two of these when forming

$$\bigvee_{i=0}^{n} T^{-i}(P).$$

Thus

$$\#(T,P,n+1\text{-names}) = \#(T,P,n\text{-names}) + 2$$

and the result follows. ∎

Example 2 *Bernoulli Processes.* We introduced Markov processes in Chapter 1, Example 3. Bernoulli processes are a special case of Markov processes. As we continue through this and the next two chapters, they will come to play an ever more critical role. Hence we will describe them in detail, and begin with almost obvious facts. Let $X = \{1,2,\ldots,s\}^{\mathbb{Z}}$, the set of all doubly infinite sequences of elements from the finite set of 'symbols' $\{1,2,\ldots,s\}$, and let

$$\boldsymbol{\pi} = \{\pi_1, \pi_2, \ldots, \pi_s\}$$

be a probability vector $\pi_i > 0$. Define μ on a 'cylinder set'

$$(i_u, i_{u+1}, \ldots, i_v) = \{x \in X \mid \text{the symbol at index } j \in (u,v) \text{ of } x \text{ is } i_j\},$$

to be

$$\pi_{i_u} \cdot \pi_{i_{u+1}} \cdot \ldots \cdot \pi_{i_v}. \tag{5.33}$$

As long as $s \neq 1$, one can easily show that using cylinder sets to form a generating tree of partitions, $(X, \mathscr{F}, \mu)$ is a non-atomic Lebesgue probability space. Define $T: X \to X$ by

$$T(\ldots, i_{-n}, \ldots, i_0, \ldots, i_n) = (\ldots, j_{-n}, \ldots, j_0, \ldots, j_n)$$

where $j_s = i_{s+1}$ (the left shift).

This is known as a Bernoulli shift, Bernoulli process, or i.i.d. (independent, identically distributed) process. A Bernoulli shift is completely specified by the probability vector $\boldsymbol{\pi}$, so often this is all that is given, i.e., the Bernoulli shift (1/2, 1/2), also called just 'the 2-shift' or Bernoulli shift (1/3, 1/3, 1/3), etc.

We want to compute $h(T)$. To do so we first need to know it is ergodic.

Lemma 5.14 *A Bernoulli shift is mixing and so ergodic (in fact it is a K-process but we're not ready to prove that yet).*

Proof Let A and B be finite unions of cylinder sets. It is clear that, once n is sufficiently large

$$\mu(T^n(A) \cap B) = \mu(A)\mu(B).$$

Such finite unions are dense in $\mathscr{F}$ (w.r.t. μ). It follows that if $T(A) = A$ then $\mu(A) = \mu(A)^2$ and so $\mu(A) = 0,1$. ■

Let P be the finite partition of X according to $i_0(x)$, the 0-position symbol, and

$$P = \{p_1, p_2, \ldots, p_s\}.$$

Sets in $\bigvee_{i=u}^{v} T^{-i}(P)$ consist of cylinders on indices (u, v), hence P clearly separates points and so generates.

We will compute

$$h(T) = h(T, P)$$

by using the Shannon–McMillan–Breiman theorem, (5.3) which identifies the entropy as the exponential shrinkage rate of a typical T,P,n-name.

Now

$$\mu(\mathbf{p}_n(x)) = \pi_{P(x)} \cdot \pi_{P(T(x))} \cdots \pi_{P(T^{n-1}(x))}$$

as $\mathbf{p}_n(x)$ is the cylinder

$$(P(x), P(T(x)), \ldots, P(T^{n-1}(x))).$$

This is then

$$\pi_1^{\sum_{i=0}^{n-1} \chi_1(T^i(x))} \times \pi_2^{\sum_{i=0}^{n-1} \chi_2(T^i(x))} \times \cdots \times \pi_s^{\sum_{i=0}^{n-1} \chi_s(T^i(x))}$$

where $\chi_k(x)$ is the characteristic function of p_k.

According to the Birkhoff theorem, given $\varepsilon > 0$, for a.e. $x \in X$, there is an N so that for $n \geq N$,

$$\frac{1}{n} \sum_{i=0}^{n-1} \chi_k(T^i(x)) = \pi_k \pm \varepsilon.$$

Therefore for large n

$$\mu(\mathbf{p}_n(x)) = (\pi_1^{\pi_1} \pi_2^{\pi_2} \cdots \pi_s^{\pi_s})^n (\pi_1 \ldots \pi_s)^{\pm \varepsilon n}.$$

For $\bar{\varepsilon} > 0$, choose ε so small that

$$-\varepsilon \log_2(\pi_1 \circ \ldots \circ \pi_s) < \bar{\varepsilon}$$

and now

$$\frac{-\log_2(\mu(\mathbf{p}_n(x)))}{n} - \sum_{i=1}^{s} \pi_i \log_2 \pi_i \pm \bar{\varepsilon}.$$

We conclude for a Bernoulli process

$$h(T) = h(T, P) = -\sum_{i=1}^{s} \pi_i \log_2 \pi_i. \tag{5.34}$$

The entropy of a Bernoulli shift is thus easily computed. With this result we can now, for example, conclude that the 2-shift and 3-shift are non-isomorphic, as they have different entropies. A much deeper fact, Ornstein's isomorphism theorem, states the converse: any two Bernoulli shifts of the same entropy *are* isomorphic. We will prove this in Chapter 7. Also notice that we have finally seen the formula $-\sum_i \pi_i \log_2 \pi_i$, which classically is the starting point for entropy theory. See Billingsley (1965) or Smorodinsky (1971) for good discussions of its abstract character. This is a generalization of our function $H(\alpha)$ which arose from Stirling's formula.

Example 3 *Ergodic Markov processes.* Let $[\pi_{i,j}]$ be the matrix of transition probabilities, and $(\pi_1, \ldots, \pi_s)$ be the stationary distribution of an ergodic Markov process. As in Chapter 1, X is the set of all doubly infinite sequences $(\ldots, i_{n-1}, i_n \ldots)$ where $\pi_{i_{n-1}, i_n} > 0$.

The measure μ is defined on cylinders $(i_0, i_1, \ldots, i_\ell)$ by

$$\mu((i_0, i_1, \ldots, i_\ell)) = \pi_{i_0} \cdot \pi_{i_0, i_1} \cdots \pi_{i_{\ell-1}, i_\ell}.$$

A generating partition can be formed by setting $P(x) = i_0(x)$ as in the previous example.

We compute $h(T) = h(T, P)$ exactly as in the i.i.d. case,

$$\begin{aligned}\mu(\mathbf{p}_n(x)) &= \pi_{P(x)} \cdot \pi_{P(x), P(T(x))} \cdots \pi_{P(T^{n-2}(x)), P(T^{n-1}(x))} \\ &= \pi_{p(x)} \prod_{\substack{i,j \\ \pi_{i,j} > 0}} \pi_{i,j}^{r_{i,j}(x)}\end{aligned}$$

where $r_{i,j}(x) = \sum_{l=0}^{n-2} \chi_{(i,j)}(T^\ell_{(x)})$ and $\chi_{(i,j)}$ is the characteristic function of the cylinder (i, j).

Once again, by the Birkhoff theorem, for $\varepsilon > 0$, for a.e. x, once n is sufficiently large

$$\mu(\mathbf{p}_n(x)) = \pi_{p(x)} \prod_{\substack{i,j \\ \pi_{i,j} > 0}} \pi_{i,j}^{(\mu(i,j) \pm \varepsilon)(n-1)}.$$

Hence for $\bar{\varepsilon} > 0$ choose ε with

$$-\varepsilon \log_2 \left(\prod_{\substack{i,j \\ \pi_{i,j} > 0}} \pi_{i,j} \right) < \bar{\varepsilon}$$

and we obtain

$$\frac{-\log_2(\mu(\mathbf{p}_n(x))}{n} = \frac{-\log_2 \pi_{p(x)}}{n} - \frac{n-1}{n}\sum_{i,j} \pi_i \pi_{i,j} \log_2 \pi_{i,j} \pm \bar{\varepsilon}.$$

The Shannon–McMillan–Breiman theorem (5.3) now implies for ergodic Markov processes

$$h(T, P) = -\sum_{i,j} \pi_i \pi_{i,j} \log_2 \pi_{i,j}. \tag{5.35}$$

Note that a Bernoulli shift is a special case of a Markov process with $\pi_{i,j} = \pi_j$ and the entropy formula of Example 3 reduces to that of Example 2.

5.7 Entropy and information from the entropy formula

As we indicated earlier, the formula for the entropy of a Bernoulli shift is the starting point for the classical development of entropy. We will now give a part of that development.

Definition 5.3 For a probability vector $\boldsymbol{\pi} = (\pi_1, \ldots, \pi_s)$ we define

$$H(\boldsymbol{\pi}) = -\sum_{i=1}^{s} \pi_i \log_2 \pi_i \tag{5.36}$$

(with the convention that extending $x \log x$ continuously to 0, $0 \log_2 0 = 0$).

This is a convex function, i.e.,

Lemma 5.15 *If $\boldsymbol{\pi}$ and $\boldsymbol{\pi}'$ are probability vectors and $0 < \lambda < 1$ then*

$$H(\lambda\boldsymbol{\pi} + (1-\lambda)\boldsymbol{\pi}') \geq \lambda H(\boldsymbol{\pi}) + (1-\lambda)H(\boldsymbol{\pi}')$$

with equality precisely when

$$\boldsymbol{\pi} = \boldsymbol{\pi}'.$$

Proof Let

$$F(\lambda) = H(\lambda\boldsymbol{\pi} + (1-\lambda)\boldsymbol{\pi}').$$

One easily computes

$$F''(\lambda) = -\sum_{i=1}^{s} \left(\frac{(\pi_i - \pi_i')^2 \ln(2)}{\lambda\pi_i + (1-\lambda)\pi_i'} \right) \leq 0$$

and equality holds only if $\pi_i = \pi_i'$ for all i. ■

Exercise 5.3 Show that as a corollary of Lemma 5.15, the maximum value of H on n-dimensional probability vectors is $\log_2(n)$.

Definition 5.4 For a finite partition P of a probability space $(X, \mathscr{F}, \mu)$, we set

$$h(P) = H((\mu(p_1), \ldots, \mu(p_s)). \tag{5.37}$$

Now for a measure-preserving transformation T of $(X, \mathscr{F}, \mu)$ we have the past algebra

$$\mathscr{P} = \bigvee_{i=-1}^{-\infty} T^{-i}(P),$$

and defined a.e., the conditional distribution of P given $\mathscr{P}$, written

$$D(P|\mathscr{P}) = (E(p_1|\mathscr{P}), E(p_2|\mathscr{P}), \ldots, E(p_s|\mathscr{P})).$$

This is a probability vector-valued function on X. We have seen earlier that

$$h(T, P) = 0$$

if and only if $D(P|\mathscr{P})$ is an elementary vector (all 0's but for a single 1) almost everywhere.

Definition 5.5 The *conditional information of P given its past $\mathscr{P}$* is

$$I(P|\mathscr{P}) = H(D(P|\mathscr{P})).$$

This is a function, not a number.

More generally, if $\mathscr{H}$ is any subalgebra of $\mathscr{F}$ then

$$D(P|\mathscr{H}) = (E(p_1|\mathscr{H}), E(p_2|\mathscr{H}), \ldots, E(p_s|\mathscr{H}))$$

and

$$I(P|\mathscr{H}) = H(D(P|\mathscr{H})).$$

This function is called the conditional information of P given $\mathscr{H}$. It is meant to measure the amount of 'information' gained when learning the set in P to which x belongs, having already known the sets in $\mathscr{H}$ to which x belong. Equivalently, it measures how much 'randomness' remains in P after having learned all of $\mathscr{H}$. These are obviously only heuristic ideas. In fact, much precision can be given them. Our intentions are more technical and less philosophical.

Set

$$\bar{h}(T, P) = \int I(P|\mathscr{P}) \, \mathrm{d}\mu. \tag{5.38}$$

Our goal is to show that

$$\bar{h}(T, P) = h(T, P). \tag{5.39}$$

First we check this equality for the examples we have considered.

Example 1

$$h(T, P) = 0 \quad \text{iff} \quad \bar{h}(T, P) = 0.$$

We have seen that $h(T, P) = 0$ iff $D(P|\mathscr{P})$ is an elementary vector a.e. But $H(\pi) = 0$ iff π is an elementary vector and the result follows.

Example 2 For an i.i.d. process, by the definition of independence,

$$D(P|\mathscr{P}) = (\pi_1, \pi_2, \ldots, \pi_s).$$

Hence $I(P|\mathscr{P}) = H(\boldsymbol{\pi})$ so

$$\bar{h}(T, P) = H(\boldsymbol{\pi}) = h(T, P).$$

Example 3 For a Markov process, $D(P|\mathscr{P}) = (\pi_{k,1}, \pi_{k,2}, \ldots, \pi_{k,s})$ if $T^{-1}(x) \in p_k$ and hence

$$\int I(P|\mathscr{P})\,\mathrm{d}\mu = \sum_{k=1}^{s} \pi_k \cdot H(\pi_{k,1}, \pi_{k,2}, \ldots, \pi_{k,s}) = h(T, P).$$

Showing that $\bar{h} = h$ can be thought of as a generalization of what happens for Markov chains. For a Markov chain all the 'information' the past algebra gives us concerning the present is contained in the single symbol at time -1.

Lemma 5.16 *Let P and H be finite partitions. Regarding H as a finite σ-algebra,*

$$h(P \vee H) = h(H) + \int I(P|H)\,\mathrm{d}\mu.$$

Proof Let $\pi_i = \mu(h_i)$ and

$$\pi_{i,j} = \frac{\mu(h_i \cap p_j)}{\mu(h_i)} = \mu(p_j|h_i).$$

Then

$$\begin{aligned} h(P \vee H) &= -\sum_{i,j} \mu(h_i \cap p_j)\log_2(\mu(h_i \cap p_j)) \\ &= -\sum_{i,j} \pi_i \pi_{i,j} \log_2(\pi_i \pi_{i,j}) \\ &= -\sum_{i,j} \pi_{i,j}\pi_i \log_2(\pi_i) - \sum_{i,j} \pi_{i,j}\pi_i \log_2(\pi_{i,j}) \\ &= \sum_i \pi_i \log_2(\pi_i) - \sum_{i,j} \pi_i \pi_{i,j} \log_2(\pi_{i,j}). \end{aligned}$$

Now

$$\int I(P|H)\,\mathrm{d}\mu = -\sum_i \sum_j \pi(p_j \cap h_i)\log_2 \frac{\mu(h_i \cap p_j)}{\mu(h_i)}$$
$$= -\sum_{i,j} \pi_{i,j}\pi_i \log_2(\pi_{i,j})$$

and

$$h(P \vee H) = h(H) + \int I(P|H)\,\mathrm{d}\mu. \qquad \blacksquare$$

Corollary 5.17 *Suppose* $\boldsymbol{\pi} = (\pi_1, \ldots, \pi_s)$ *and* $\boldsymbol{\pi}' = (\pi'_1, \ldots, \pi'_t)$ *are probability vectors and, for* $0 < \lambda < 1$,

$$\boldsymbol{\pi}'' = (\lambda\pi_1, \lambda\pi_2, \ldots, \lambda\pi_s, (1-\lambda)\pi'_1, \ldots, (1-\lambda)\pi'_s).$$

Then

$$H(\pi'') = \lambda H(\pi) + (1-\lambda)H(\pi') + H(\lambda).$$

Proof Exercise 5.4. ■

Exercise 5.4 Prove Corollary 5.17.

Lemma 5.18

$$h(T, P) = \lim_{n\to\infty} \frac{1}{n} h\left(\bigvee_{i=0}^{n-1} T^{-i}(P)\right).$$

Proof Let $\varepsilon > 0$ and select N so that for all $n \geq N$, for a set G of all but ε of the $x \in X$ we have

$$\mu(\mathbf{p}_n(x)) = 2^{-(h(T,P)\pm\varepsilon)n}.$$

Let $B = G^c$, $\mu(B) < \varepsilon$. Consider the probability vectors

$$\boldsymbol{\pi} = (\pi_1, \pi_2, \ldots, \pi_m) = D\left(\bigvee_{i=0}^{n-1} T^{-i}(P)|B\right)$$

and

$$\boldsymbol{\pi}' = (\pi'_1, \ldots, \pi'_t) = D\left(\bigvee_{i=0}^{n-1} T^{-i}(P)|G\right).$$

Now

$$h\left(\bigvee_{i=0}^{n-1} T^{-i}(P)\right) = \mu(B)H(\boldsymbol{\pi}) + \mu(G)H(\boldsymbol{\pi}') + H(\mu(B))$$

by Corollary 5.17.

If ε is sufficiently small,

$$H(\varepsilon) = -\varepsilon \log_2(\varepsilon) - (1 - \varepsilon)\log_2(1 - \varepsilon) < \bar{\varepsilon}$$

and

$$h\left(\bigvee_{i=0}^{n-1} T^{-i}(P)\right) = (1 \pm \varepsilon)H(\pi') \pm \varepsilon H(\pi) \pm \bar{\varepsilon}. \tag{5.40}$$

Now

$$\begin{aligned} H(\pi') &= -\sum \pi_i' \log_2 \pi_i' \\ &= -\sum \pi_i' \log_2\left(\frac{2^{-(h(T,P)\pm\varepsilon)n}}{\mu(G)}\right) \\ &= -\sum \pi_i'\{-(h(T,P) \pm \varepsilon)n) - \log_2(\mu(G))\} \\ &= n(h(T,P) \pm \varepsilon) + \log_2(\mu(\mathrm{G})), \end{aligned}$$

and

$$H(\boldsymbol{\pi}) \le n \log_2 s$$

as $\boldsymbol{\pi}$ has at most s^n elements and H is maximized when they are all of equal size (Exercise 5.3). Thus (5.40) gives

$$\frac{1}{n}\left(\bigvee_{i=0}^{n-1} T^{-i}(P)\right) = (1 \pm \varepsilon)(h(T,P) \pm \varepsilon) - \frac{\log_2(\mu(G))}{n} \pm \varepsilon \log_2(s) \pm \bar{\varepsilon}.$$

Letting $n \to \infty$, forces both ε and $\bar{\varepsilon}$ to zero and we get the result. ∎

Theorem 5.19 *Let P be a finite partition of X and $\mathscr{P}$ its past algebra.*

$$h(T,P) = \int I(P|\mathscr{P})\,\mathrm{d}\mu = \bar{h}(T,P).$$

Proof By iterative applications of

$$h(P \vee H) = h(H) + \int I(P|H),$$

we have

$$\begin{aligned} h\left(\bigvee_{i=0}^{n-1} T^{-i}(P)\right) &= \int I(T^{-n+1}(P))\,\mathrm{d}\mu + \int I(T^{-n+2}(P)|T^{-n+1}(P))\,\mathrm{d}\mu \\ &\quad + \int I(T^{-n+3}(P)|T^{-n+1}(P) \vee T^{-n+2}(P))\,\mathrm{d}\mu \\ &\quad + \cdots + \int I\left(P\middle|\bigvee_{i=-1}^{-n+1} T^{-i}(P)\right)\mathrm{d}\mu \\ &= \int I(P)\,\mathrm{d}\mu + \int I(P|T(P))\,\mathrm{d}\mu + \cdots \int I\left(P\middle|\bigvee_{i=0}^{-n+1} T^{-i}(P)\right)\mathrm{d}\mu. \end{aligned}$$

Setting

$$f_j = I\left(P \,\middle|\, \bigvee_{i=-1}^{-j} T^{-i}(P)\right)$$

we know

$$f_j \to I(P|\mathscr{P})$$

pointwise and in L^1, by the L^1-martingale theorem, $(D(P|\bigvee_{i=1}^{j} T^i(P))$ form the martingale.

Hence

$$\begin{aligned} h(T, P) &= \lim_{n\to\infty} \frac{1}{n} h\left(\bigvee_{i=0}^{n-1} T^{-i}(P)\right) \\ &= \lim_{n\to\infty} \frac{1}{n} \sum_{i=0}^{n-1} \int f_i \,\mathrm{d}\mu \\ &= \int I(P|\mathscr{P})\,\mathrm{d}\mu. \end{aligned}$$

■

Exercise 5.5 Within this discussion, there arises a natural notion of the conditional entropy of a partition conditioned on a factor algebra $\mathscr{A}$.

1. $h(P|\mathscr{A}) = \int I(P|\mathscr{A})\,\mathrm{d}\mu$. For example, $h(T, P) = h(P|\mathscr{P}_p)$.

This extends to a natural notion of the conditional entropy of a process, conditioned on a factor algebra.

2. $h(T, P|\mathscr{A}) = \int I(P|\mathscr{P}_p \vee \mathscr{A}) = \lim_{n\to\infty} \frac{1}{n} h\left(\bigvee_{i=0}^{n-1} T^{-i}(P)|\mathscr{A}\right)$.

   (a) The first equal sign in 2 is a definition, of course. Prove the second one.

   (b) If the partition H is a generator for $\mathscr{A}$, then following the argument for Lemma 5.16, show

$$h(T, P \vee H) = h(T, H) + h\left(T, P \,\middle|\, \bigvee_{i=-\infty}^{\infty} T^{-i}(H)\right).$$

5.8 More about zero entropy and tail fields

Returning to the case of zero entropy transformations, remember we know $h(T, P) = 0$ iff

$$P \subset \mathscr{P} = \bigvee_{i=-1}^{-\infty} T^{-i}(P).$$

This, of course, says

$$T^{-j}(P) \subset \bigvee_{i=-j-1}^{-\infty} T^{-i}(P)$$

and so

$$\mathscr{P} \subset T^{-1}(\mathscr{P}).$$

If we define

$$\mathscr{T}_P = \bigcap_{j=-1}^{-\infty} \left(\bigvee_{i=j}^{-\infty} T^{-i}(P) \right),$$

the *tail field* of P, then we obtain the following result.

Corollary 5.20

$$h(T, P) = 0 \quad \textit{iff} \quad P \subset \mathscr{T}_P. \qquad \blacksquare$$

We want to generalize this last corollary to the following theorem (we have taken this argument directly from Smorodinsky (1971).

Theorem 5.21 *If P and Q are finite partitions and $P \subset \mathscr{T}_Q$, then $h(T, P) = 0$.*

Lemma 5.22

$$h\left(T^k, \bigvee_{i=0}^{k-1} T^{-i}(P)\right) = kh(T, P)$$

Proof Exercise 5.6.

Exercise 5.6 Prove Lemma 5.22.

Corollary 5.23 $h(T^k) = kh(T)$.

Proof of Theorem 5.21

$$h\left(T^k, \bigvee_{i=0}^{k-1} T^{-i}(P)\right) = \int I\left(\bigvee_{j=0}^{k-1} T^{-j}(P) \middle| \bigvee_{i=-1}^{-\infty} (T^k)^i \left(\bigvee_{j=0}^{k-1} T^{-j}(P)\right)\right)$$

$$= \int I\left(\bigvee_{j=0}^{k-1} T^{-j}(P) \middle| \bigvee_{i=-1}^{-\infty} T^{-j}(P)\right).$$

By lemma 5.22

$$h(T, P) = \frac{1}{k} \int I\left(\bigvee_{j=0}^{k-1} T^{-j}(P) \middle| \bigvee_{j=-1}^{-\infty} T^{-j}(P)\right)$$

$$= \lim_{k\to\infty} \frac{1}{k} \int I\left(\bigvee_{j=0}^{k-1} T^{-j}(P) \middle| \bigvee_{j=-1}^{-\infty} T^{-j}(P)\right).$$

By lemma 5.8

$$h(T, P) = \lim_{k\to\infty} \frac{1}{k} \int I\left(\bigvee_{j=0}^{k-1} T^{-j}(P)\right).$$

Consider:

$$(1) \qquad \frac{1}{k} \int I\left(\bigvee_{j=0}^{k-1} T^{-j}(P \vee Q) \middle| \bigvee_{j=-1}^{-\infty} T^{-j}(P \vee Q)\right)$$
$$= \frac{1}{k} \int I\left(\bigvee_{j=0}^{k-1} T^{-j}(P) \middle| \bigvee_{j=-1}^{-\infty} T^{-j}(P \vee Q)\right)$$
$$+ \frac{1}{k} \int I\left(\bigvee_{j=0}^{k-1} T^{-j}(Q) \middle| \bigvee_{j=-\infty}^{k-1} T^{-j}(P) \vee \bigvee_{j=-\infty}^{-1} T^{-j}(Q)\right)$$

and

$$(2) \qquad \frac{1}{k} \int I\left(\bigvee_{j=0}^{k-1} T^{-j}(P \vee Q) \middle| \bigvee_{j=-1}^{-\infty} T^{-j}(P)\right)$$
$$= \frac{1}{k} \int I\left(\bigvee_{j=0}^{k-1} T^{-j}(P) \middle| \bigvee_{j=-1}^{-\infty} T^{-j}(P)\right)$$
$$+ \frac{1}{k} \int I\left(\bigvee_{j=0}^{k-1} T^{-j}(Q) \middle| \bigvee_{j=-\infty}^{k-1} T^{-j}(P)\right).$$

The left hand side of (2) is less than or equal to

$$\frac{1}{k} \int I\left(\bigvee_{j=0}^{k-1} T^{-j}(P \vee Q)\right)$$

and greater than or equal to the left side of (1), both of which tend to $h(T, P \vee Q)$ as $k \to \infty$.

Each term of the right hand side of (2) is less than or equal to the corresponding term on the right hand side of (1). Thus

$$\frac{1}{k} \int I\left(\bigvee_{j=0}^{k-1} T^{-j}(P) \middle| \bigvee_{j=-1}^{-\infty} T^{-j}(P)\right) - \frac{1}{k} \int I\left(\bigvee_{j=0}^{k-1} T^{-j}(P) \middle| \bigvee_{j=-1}^{-\infty} T^{-j}(P \vee Q)\right)$$

must converge to 0 as $k \to \infty$, i.e.

$$\lim_{k\to\infty} \frac{1}{k} \int I\left(\bigvee_{j=0}^{k-1} T^{-j}(P) \middle| \bigvee_{j=-1}^{-\infty} T^{-j}(P \vee Q)\right) = h(T, P). \tag{5.41}$$

If $P \subseteq \mathscr{T}_Q$, then for all $j \geq 0$,

$$T^{-j}(P) \subseteq \bigvee_{j=-1}^{-\infty} T^{-j}(Q) \subseteq \bigvee_{j=-1}^{-\infty} T^{-j}(P \vee Q)$$

and

$$\bigvee_{j=0}^{k-1} T^{-j}(P) \subseteq \bigvee_{j=-1}^{-\infty} T^{-j}(P \vee Q)$$

and so

$$\frac{1}{k}\int I\left(\bigvee_{j=0}^{k-1} T^{-j}(P)\middle|\bigvee_{j=-1}^{-\infty} T^{-j}(P \vee Q)\right) = 0$$

and hence by (5.6), $h(T, P) = 0$. ∎

Exercise 5.7

1. Show that if $h(T, P) = 0$ and $h(T, Q) = 0$ then $H(T, P \vee Q) = 0$.
2. Use part 1 to show that for any transformation T there is a maximal σ-algebra π, called the Pinsker algebra, so that for any $P \subset \pi$, $h(T, P) = 0$.
3. Show that as irrational rotations and permutations of finite sets have zero entropy, any minimal isometry has zero entropy. Hint: spectrum.

5.9 Even more about the K-property

Corollary 5.24 *$h(T, P) \neq 0$ for all non-trivial P iff $\mathcal{T}_p$ is trivial for all P. (Trivial means consists only of sets of measure 0 or 1.)* ∎

We want to see that this is also equivalent to $\mathcal{T}_P$-trivial for a single generating partition P.

Theorem 5.25 *If P and Q are finite partitions, $Q \subset \bigvee_{i=-\infty}^{+\infty} T^{-j}(P)$ then $\mathcal{T}_Q \subseteq \mathcal{T}_P$.*

We first deal with some preliminaries. Note that if $Q \subset \bigvee_{i=-k}^{k} T^{-i}(P)$ then $\mathcal{T}_Q \subset \mathcal{T}_P$ is easy to show.

Next notice that if $f, g \in L^1$ and $\mathscr{A}$ is any algebra,

$$\|E(f|\mathscr{A}) - E(g|\mathscr{A})\|_1 \leq \|f - g\|_1.$$

It follows that given any $s \in \mathbb{Z}^+$ and $\varepsilon > 0$, there is a δ so that if P and P' are partitions with the same state space and

$$\mu(P \,\Delta\, P') < \delta$$

then

$$\int |I(P|\mathscr{A}) - I(P'|\mathscr{A})| \,\mathrm{d}\mu < \varepsilon$$

for any algebra $\mathscr{A}$.

Remember from Lemma 5.15 that $H(\pi)$ is a strictly convex function, i.e.,

$$H(\sum a_i \pi_i) \geq \sum a_i H(\pi_i)$$

where $\boldsymbol{a}$ is a strictly positive probability vector, equality holding exactly when all π_i are equal. The following is just a restatement of this.

Corollary 5.26 *If P and Q are finite partitions then*

$$\int I(P|Q)\,\mathrm{d}\mu \leq h(P)$$

and equality holds iff P is independent of Q (written $P \perp Q$), i.e., $D(P|Q) = D(P)$ a.s.

This leads to the following strengthening.

Theorem 5.27 *For any algebras $\mathscr{A}_1$ and $\mathscr{A}_2$, and finite partition P,*

$$\int I(P|\mathscr{A}_1 \vee \mathscr{A}_2)\,\mathrm{d}\mu \leq \int I(P|\mathscr{A}_1)\,\mathrm{d}\mu$$

and equality holds iff

$$D(P|\mathscr{A}_1 \vee \mathscr{A}_2) = D(P|\mathscr{A}_1) \quad \text{a.s.}$$

Proof We begin by supposing $\mathscr{A}_1$ is generated by a finite partition $Q = \{q_1, \ldots, q_s\}$. Now

$$D(P|\mathscr{A}_2) = \sum_{q_i} E(q_i|\mathscr{A}_2)\left(\frac{D(P \cap q_i|\mathscr{A}_2)}{E(q_i|\mathscr{A}_2)}\right)$$

and

$$D(P|Q \vee \mathscr{A}_2) = \sum_{q_i} \chi_{q_i}\left(\frac{D(P \cap q_i|\mathscr{A}_2)}{E(q_i|\mathscr{A}_2)}\right)$$

a.s. where

$$\frac{D(P \cap q_i|\mathscr{A}_2)}{E(q_i|\mathscr{A}_2)} = \frac{(E(p_1 \cap q_i|\mathscr{A}_2), (E(p_2 \cap q_i|\mathscr{A}_2), \ldots)}{E(q_i|\mathscr{A}_2)}$$

is a probability vector-valued function. For a.e. $x \in X$,

$$\begin{aligned} I(P|\mathscr{A}_2)(x) &= H(D(P|\mathscr{A}_2)(x) \\ &\geq \sum_{q_i} E(q_i|\mathscr{A}_2)(x) H\left(\frac{D(P \cap q_i|\mathscr{A}_2)(x)}{E(q_i|\mathscr{A}_2)(x)}\right), \end{aligned}$$

equality holding iff

$$\frac{D(P \cap q_i|\mathscr{A}_2)(x)}{E(q_i|\mathscr{A}_2)(x)} = D(P|\mathscr{A}_2)(x)$$

by Corollary 5.26. Thus

$$\int_X I(P|\mathscr{A}_2)\,\mathrm{d}\mu \geq \sum_{q_i} \int_X E(q_i|\mathscr{A}_2) H\left(\frac{D(P \cap q_i|\mathscr{A}_2)}{E(q_i|\mathscr{A}_2)}\right) \mathrm{d}\mu. \tag{5.42}$$

As $H(D(P \cap q_i|\mathscr{A}_2)/E(q_i|\mathscr{A}_2))$ is $\mathscr{A}_2$ measurable, the left hand side of (5.42) is

$$\sum_{q_i} \int_X E \int \left(\chi_{q_i}\left(\frac{D(P \cap q_i|\mathscr{A}_2)}{E(q_i|\mathscr{A}_2)}\right)\right) \mathrm{d}\mu = \int_X \sum_{q_i} \chi_{q_i} I\left(\frac{d(P \cap q_i|\mathscr{A}_2)}{E(q_i|\mathscr{A}_2)}\right) \mathrm{d}\mu$$

$$= \int_X I(D(P|Q \vee \mathscr{A}_2))\,\mathrm{d}\mu.$$

Equality in (5.42) holds iff $D(P \cap q_i|\mathscr{A}_2)/E(q_i|\mathscr{A}_2) = D(P|\mathscr{A}_2)$ for all q_i for a.e. x, i.e.,

$$D(P|Q \vee \mathscr{A}_2) = D(P|\mathscr{A}_2) \quad \text{a.e.}$$

Letting Q_i refine down to $\mathscr{A}_1$,

$$\int I(P|Q_{i+1} \vee \mathscr{A}_2)\,\mathrm{d}\mu \leq \int I(P|Q_i \vee \mathscr{A}_2)\,\mathrm{d}\mu \tag{5.43}$$

by what we just proved and taking limits

$$\int I(P|\mathscr{A}_2 \vee \mathscr{A}_2)\,\mathrm{d}\mu \leq \int I(P|\mathscr{A}_1)\,\mathrm{d}\mu.$$

If equality holds here it holds in (5.43) for all i and

$$D(P|Q_i \vee \mathscr{A}_2) = D(P|\mathscr{A}_2) \quad \text{a.e.}$$

for all i and $D(P|\mathscr{A}_1 \vee \mathscr{A}_2) = D(P|\mathscr{A}_2)$. ∎

Corollary 5.28 *For any finite partitions P and Q,*

$$D(P|\mathscr{P}_P \vee \mathscr{T}_Q) = D(P|\mathscr{P}_P).$$

Proof We saw earlier that $1/k \int I(\bigvee_{j=0}^{k-1} T^{-j}(P)|\bigvee_{j=-1}^{-\infty} T^{-j}(P \vee Q))\,\mathrm{d}\mu$ converges in k to $h(T, P)$. This quantity can be written

$$\frac{1}{k}\sum_{j=0}^{k-1} \int I\left(T^{-j}(P) \middle| \bigvee_{i=j-1}^{-\infty} T^{-i}(P) \vee \bigvee_{i=-1}^{-\infty} T^{-i}(Q)\right) \mathrm{d}\mu$$

$$= \frac{1}{k}\sum_{j=0}^{k-1} \int I\left(P \middle| \bigvee_{i=-1}^{-\infty} T^{-i}(P) \vee \bigvee_{i=-j-1}^{-\infty} T^{-i}(Q)\right) \mathrm{d}\mu.$$

This is the Césaro average of the sequence

$$\alpha_j = \int I\left(P \,\middle|\, \bigvee_{i=-1}^{-\infty} T^{-i}(P) \vee \bigvee_{i=-j-1}^{-\infty} T^{-i}(Q)\right) d\mu,$$

an increasing bounded sequence, hence α_j converges to $h(T, P)$. But now

$$\int I(P|\mathscr{P}_P)\,d\mu \geq \int I(P|\mathscr{P}_P \vee \mathscr{T}_Q)\,d\mu \geq \alpha_j$$

for all j hence

$$\int I(P|\mathscr{P}_P \vee \mathscr{T}_Q)\,d\mu = \int I(P|\mathscr{P}_P)\,d\mu$$

and so

$$D(P|\mathscr{P}_P \vee \mathscr{T}_Q) = D(P|\mathscr{P}_P). \qquad \blacksquare$$

We need one last ingredient, a reverse martingale theorem.

Lemma 5.29 *Let $G_1 \subset G_2 \subset \cdots \subset G_n$ be a finite sequence of σ-algebras and $f \in L^1(\mu)$, $\lambda \in R$ be fixed. Set*

$$M = \left\{x : \max_{1 \leq k \leq n} E(f|G_k) \geq \lambda\right\}.$$

For any set $A \in G_1$,

$$\int_{A \cap M} f\,d\mu \geq \lambda\mu(M \cap A).$$

Proof Let $M_k = \{x : E(f|G_i) < \lambda$ for $i < k$ but $E(f|G_k) \geq \lambda\}$. The M_k, $k = 1, 2, \ldots, n$ are disjoint and cover M. Each M_k is G_k measurable.

$$\begin{aligned}\int_{M \cap A} f\,d\mu &= \sum_{k=1}^{n} \int_{M_k \cap A} f\,d\mu \\ &= \sum_{k=1}^{n} \int_{M_k \cap A} E(f|G_k)\,d\mu \\ &\geq \sum_{k=1}^{n} \lambda\mu(M_k \cap A) = \lambda\mu(M \cap A).\end{aligned}$$

Theorem 5.30 (Reverse martingale theorem, Doob 1953) *If $\{G_1 \supseteq G_2 \supseteq \cdots\}$ is a sequence of σ-algebras which decrease to G and if $f \in L'(\mu)$, then $E(f|G_k)$ converges pointwise and in L^1 to $E(f|G)$.*

Proof L^1 convergence will follow from pointwise convergence and uniform integrability.

Let

$$\begin{aligned} A &= A(\lambda_1, \lambda_2) \\ &= \{x : \underline{\lim}\, E(f|G_n) < \lambda_1 < \lambda_2 < \overline{\lim}\, E(F|G_n)\}. \end{aligned}$$

If we show $\mu(A) = 0$ for all λ_1, λ_2 then we will have pointwise convergence.

Let $M_n = \{x : \max_{1 \le k \le n} E(f|G_k) \ge \lambda_2\}$. By lemma 5.29

$$\int_{M_n \cap A} f \, d\mu \ge \lambda_2 \mu(M_n \cap A).$$

As $n \to \infty$, $M_n \cap A \to A$ and so

$$\int_A f \, d\mu \ge \lambda_2 \mu(A).$$

Replacing f by $-f$ and λ_2 by $-\lambda_1$, we get similarly

$$\int_A f \, d\mu \le \lambda_1 \mu(A).$$

Hence $\mu(A) = 0$.

To verify uniform integrability, assume $f \ge 0$. Notice that $h_k(\delta) = \sup\{\int_A E(f|G_k)\, d\mu | \mu(A) < \delta\}$ is actually a maximum and is achieved on a set A which is G_k measurable.

Hence $h_k(\delta) \le h_{k-1}(\delta)$ is decreasing in k. Thus if δ is such that for any A, $\mu(A) < \delta$, $\int_A f \, d\mu < \varepsilon$, then for all G_k and $\mu(A) < \delta$,

$$\int_A E(f|G_k)\, d\mu < \delta.$$

This shows $E(f|G_k)$ converges in L^1 and pointwise to some function f^*. Now f^* is G measurable so all we need show is that for $A \in G$, $\int_A f^* \, d\mu = \int_A f \, d\mu$ to conclude $f^* = E(f|G)$.

But $\int_A f^* \, d\mu = \lim_{k\to\infty} \int_A E(f|G_k)\, d\mu$, and as $A \in G \subseteq G_k$, this is equal to $\int_A f \, d\mu$ and we are done. ■

Theorem 5.30, for example, tells us that for finite partitions P and Q, $D(Q|\bigvee_{i=-j}^{-\infty} T^{-i}(P))$ converges pointwise and in L^1 to $D(Q|\mathscr{T}(P))$ and so $\int I(Q|\bigvee_{i=-j}^{-\infty} T^{-i}(P))\, d\mu$ converges to $\int I(Q|\mathscr{T}_P)\, d\mu$.

Proof of Theorem 5.25 We prove that for $Q \subset \bigvee_{i=-\infty}^{\infty} T^{-i}(P)$ we must have $\mathscr{T}_Q \subseteq \mathscr{T}_P$ by showing that any $R \subset \mathscr{T}_Q$ has conditional entropy 0 with respect to $\mathscr{T}_P$ and hence is $\mathscr{T}_P$ measurable. Now using Corollary 5.28,

$$D\left(\bigvee_{i=-k}^{k} T^{-i}(P) \middle| \bigvee_{i=-k-1}^{-\infty} T^{-i}(P) \vee \mathscr{T}_Q\right) = D\left(\bigvee_{i=-k}^{k} T^{-i}(P) \middle| \bigvee_{i=-k-1}^{-\infty} T^{-i}(P)\right) \quad \text{a.s.,}$$

so

$$D\left(\bigvee_{i=-k}^{k} T^{-i}(P)\,\middle|\,\bigvee_{i=-k-1}^{-\infty} T^{-i}(P) \vee R\right) = D\left(\bigvee_{i=-k}^{k} T^{-i}(P)\,\middle|\,\bigvee_{i=-k-1}^{-\infty} T^{-i}(P)\right).$$

Let $k > m$ and $S \subset \bigvee_{i=-m}^{m} T^{-i}(P)$. Then

$$D\left(S\,\middle|\,\bigvee_{i=-k-1}^{-\infty} T^{-i}(P)\right) = D\left(S\,\middle|\,\bigvee_{i=-k-1}^{-\infty} T^{-i}(P) \vee R\right)$$

and so

$$\int I\left(S\,\middle|\,\bigvee_{i=-k-1}^{-\infty} T^{-i}(P)\right) \mathrm{d}\mu = \int I\left(S\,\middle|\,\bigvee_{i=-k-1}^{-\infty} T^{-i}(P) \vee R\right).$$

Letting $k \to \infty$, using the reverse martingale theorem

$$\int I(S|\mathscr{T}_p)\,\mathrm{d}\mu = \int I(S|\mathscr{T}_p \vee R)\,\mathrm{d}\mu.$$

This holds for any S lying in a finite span of the $T^{-j}(P)$.

As $Q \subset \bigvee_{i=-\infty}^{\infty} T^{-i}(P)$ and $R \subset \mathscr{T}_Q$, $R \subset \bigvee_{i=-\infty}^{\infty} T^{-i}(P)$ so there is a sequence of partitions $S_i \subset \bigvee_{j=-m(i)}^{m(i)} T^{-j}(P)$ and $S_i \to R$ in symmetric difference. Thus

$$\int I(S_i|\mathscr{T}_p)\,\mathrm{d}\mu \to \int I(R|\mathscr{T}_p)\,\mathrm{d}\mu,$$

and

$$\int I(S_i|\mathscr{T}_p)\,\mathrm{d}\mu = \int I(S_i|\mathscr{T}_p \vee R)\,\mathrm{d}\mu \to \int I(R|\mathscr{T}_p \vee R)\,\mathrm{d}\mu = 0.$$

We conclude $I(R|\mathscr{T}_p) = 0$ a.e. and so $R \subset \mathscr{T}_p$ and $\mathscr{T}_Q \subseteq \mathscr{T}_P$. ■

Corollary 5.31 *If $\mathscr{T}_p$ is trivial for a generating partition P then $\mathscr{T}_Q$ is trivial for all Q.* ■

This now gives us a circle of facts about the K-property analogous to Proposition 4.19 on weakly mixing.

Definition 5.32 *The following are all equivalent:*

(1) *$\mathscr{T}_P$ is trivial for all finite P;*

(2) *$h(T, P) \neq 0$ for all finite P;*

(3) *$D(P|\bigvee_{i=-j}^{-\infty} T^{-i}(P)) \underset{j}{\rightarrow} D(P)$ for all finite P;*

(4) *T is a K-automorphism.*

If T has a finite generating partition P, these are also equivalent to

(5) $\mathcal{T}_p$ *is trivial;*

(6) $D(\bigvee_{i=-k}^{k} T^{-i}(P) | \bigvee_{i=-j}^{-\infty} T^{-i}(P)) \to D(\bigvee_{i=-k}^{k} T^{-i}(P))$ *as* $j \to \infty$.

Proof We already know the equivalence of (1), (2) and (5) and that (4) → (2). That (3) → (6) and (3) → (4) are direct. That (1) → (3) is the reverse martingale theorem. To complete the proof we show (6) → (5) and by the same reasoning (3) → (2).

Suppose $A \in \mathcal{T}_p$, $\mu(A) \neq 0, 1$. Let k be chosen and $A' \in \bigvee_{i=-k}^{k} T^{-i}(P)$ so that $|\mu(A|A') - \mu(A)| = |\mu(A \cap A')/\mu(A') - \mu(A)| > 0$. But by (5), $A \perp A'$, i.e., $\mu(A|A') = \mu(A)$ contradicting $\mu(A) \neq 0, 1$. ■

The 'usual' definitions of a K-automorphism are conditions (1), (2) or (6).

Exercise 5.8 Show that a mixing Markov chain is a K-automorphism.

Exercise 5.9 Show that a rank-1 cutting and stacking construction always has zero entropy.

5.10 Entropy for non-ergodic maps

We end this chapter with a few remarks concerning the entropy theory for not necessarily ergodic maps.

Working with counts of names one can show without ergodicity that

$$h(T, P, \varepsilon) = \lim_{n \to \infty} h(T, P, \varepsilon, n)$$

as given in Definition 5.1, is in fact a limit and as a function of ε is monotone non-increasing.

One can show

$$\lim_{\varepsilon \to 0} h(T, P, \varepsilon) = \operatorname*{ess\text{-}sup}_{y \in Y} (h(T_y, P)) \tag{5.44}$$

where T_y, $y \in Y$ are the ergodic components of T. This is one possible choice for $h(T, P)$, but certainly not the best.

A Shannon–McMillan–Breiman theorem also holds. For μ-a.e. x

$$\lim_{n \to \infty} -\frac{\log_2(\mu(\mathbf{p}_n(x)))}{n} = h(T_y, P) \tag{5.45}$$

where T_y is the ergodic component containing x.

Using the classical entropy function, one can also compute

$$\lim_{n \to \infty} \frac{1}{n} h\left(\bigvee_{i=0}^{n-1} T^{-i}(P) \right) = h(P|\mathscr{P}) = \int_Y h(T_y, P)\, dm. \tag{5.46}$$

Classical proofs of these results can be found in Jacobs (1962) and Krengel (1985). Working from the ergodic decomposition of Chapter 2, the techniques of this chapter can also be extended to their proof. We leave this as a challenge to the truly motivated reader.

We will take as our definition

$$h(T, P) = \lim_{n\to\infty} \frac{1}{n} h\left(\bigvee_{i=0}^{n-1} T^{-i}(P) \right) = h(P|\mathscr{P}) = \int I(P|\mathscr{P})\, d\mu.$$

All the equalities here are either definitions or provable without ergodicity (see Theorem 5.19). We will use these rather extensively in Chapter 7, but will only use the Shannon–McMillan–Breiman theorem for ergodic systems.

6 Joinings and disjointness

The fundamental question, and certainly the question behind all our work in Chapters 4 and 5, has been how to tell whether two dynamical systems are isomorphic. A more refined question has been, what structures exist within a dynamical system that can be used to distinguish it from others? Entropy and the point spectrum are two such structures which we have investigated rather carefully. Also mixing properties are of this sort. In this chapter we will approach our original question from a slightly different perspective.

Given two dynamical systems, how closely can they be matched to one another? What we will consider are the ways of joining two systems together as factors of some common system. As we shall see, the space of all such joinings will carry a compact metric topology, and within this space we can search for isomorphisms, or obstacles to them.

We will see that a number of mixing properties can be characterized in terms of these spaces of joinings. Further, the factor algebras and centralizer of a system are governed by the self-joinings. In this context we will construct some interesting examples. All this leads up to Chapter 7 where we prove Krieger's representation theorem and Ornstein's isomorphism theorem through careful construction and analyses within certain spaces of joinings.

6.1 Joinings

Let $\bar{X} = (X, \mathscr{F}, \mu, T)$ and $\bar{Y} = (Y, \mathscr{G}, v, S)$ be two ergodic dynamical systems.

Definition 6.1 A *joining* of $\bar{X}$ and $\bar{Y}$ is a $T \times S$ invariant measure $\hat{\mu}$ on $X \times Y$ for which all sets of the form $A \times B$, $A \in \mathscr{F}$, $B \in \mathscr{G}$ are measurable and whose *marginal* measures

$$\hat{\mu}(A \times Y) = \mu(A)$$

and

$$\hat{\mu}(X \times B) = v(B),$$

i.e., agree with μ and v.

We assume the $\hat{\mu}$ measurable sets to be the completion of $\mathscr{F} \times \mathscr{G}$.

Theorem 6.1 *Suppose $\bar{X}$ and $\bar{Y}$ are dynamical systems on Lebesgue spaces, and $\{A_i\} \subset \mathscr{F}$ and $\{B_i\} \subset \mathscr{G}$ are countable, T- and S-invariant generating* *subalgebras respectively. Suppose $\bar{\mu}_0$ is a $T \times S$-invariant additive set function on sets of the form $A_i \times B_j$ satisfying*

$$\bar{\mu}_0(A_i \times Y) = \mu(A_i)$$

and

$$\bar{\mu}_0(X \times B_j) = \nu(B_j).$$

Then $\bar{\mu}_0$ extends to a unique joining $\bar{\mu}$ of the two systems and the measure space $(X \times Y, \mathscr{F} \times \mathscr{G}, \bar{\mu})$ is automatically Lebesgue.

Proof Using sets $A_i \times B_j$ we build a refining and generating tree of partitions $\{P_i\}$,

$$P_i = \bigvee_{j=1}^{i} (A_j \times B_j, A_j \times B_j^c, A_j^c \times B_j, A_j^c \times B_j^c).$$

Of course, setting

$$Q_i = \bigvee_{j=1}^{i} (A_j, A_j^c), \quad H_i = \bigvee_{j=1}^{i} (B_j, B_j^c),$$

these are refining and generating in X and Y separately and

$$P_i = (Q_i \times Y) \vee (X \times H_i).$$

The additive set function μ_0 is defined on our tree $\{P_i\}$, and is $T \times S$-invariant. All we need to show is that the empty chains have measure 0.

Suppose $\mathscr{C} = \{C_i\}$ is a chain in $\{P_i\}$. There are then two chains $\mathscr{C}' = \{C_i'\}$ in $\{Q_i\}$ and $\mathscr{C}'' = \{C_i''\}$ in $\{H_1\}$ with $C_i = C_i' \times C_i''$.

Now $\mathscr{C}$ is an empty chain if and only if one of C' or C'' is empty. As $\bar{X}$ and $\bar{Y}$ are Lebesgue, the empty chains in $\{Q_i\}$ and in $\{H_i\}$ have respectively μ and ν measure 0. As $\bar{\mu}_0$ agrees with μ and ν on its marginals, the $\{P_i\}$-empty chains have $\bar{\mu}_0$ measure 0. ∎

This result will make joinings relatively easy to construct. All we need do is build $\bar{\mu}_0$ on a tree. More importantly, that there is a 1–1 correspondence between additive set functions with the proper marginals and joinings will allow us to topologize the space of joinings.

We have defined a joining as a measure on $X \times Y$. Such arise any time both $\bar{X}$ and $\bar{Y}$ are embedded as factors of some common supersystem $\bar{Z} = (Z, \mathscr{B}, \eta, U)$. If $\varphi_1 : Z \to X$ and $\varphi_2 : Z \to Y$ are factor maps, then restricting η to the smallest σ-algebra containing $\varphi_1^{-1}(\mathscr{F}) \vee \varphi_2^{-1}(\mathscr{G})$, we get a version of a joining. To put η on $X \times Y$, define

$$\bar{\mu}_{\bar{Z}}(A \times B) = \eta(\varphi_1^{-1}(A) \cap \varphi_2^{-1}(B)).$$

That this is an additive set function is easy to check. By Theorem 6.1, it extends to a joining.

The factor of $\bar{Z}$ generated by $\varphi_1^{-1}(\mathscr{F}) \vee \varphi_2^{-1}(\mathscr{G})$ is isomorphic to

$$(X \times Y, \mathscr{F} \times \mathscr{G}, \bar{\mu}_{\bar{Z}}, T \times S).$$

Quite often we will construct supersystems like $\bar{Z}$ and refer to them as joinings of some pair of factors. It is in the sense we have just described that the word joining is intended.

Definition 6.2 By $J(\bar{X}, \bar{Y})$ we mean the space of all joinings of $\bar{X}$ and $\bar{Y}$.

$J(\bar{X}, \bar{Y})$ is never empty as it always contains $\mu \times \nu$. It is convex, since if $\bar{\mu}_1$ and $\bar{\mu}_2$ are both joinings, then so is $\alpha\bar{\mu}_1 + (1 - \alpha)\bar{\mu}_2, 0 \leq \alpha \leq 1$.

We want to topologize $J(\bar{X}, \bar{Y})$; we give an explicit metric. For a generating tree $\{P_i\}$ as constructed above let

$$\|\bar{\mu}_1(P_i), \bar{\mu}_2(P_i)\| = \frac{1}{2} \sum_{p \in P_i} |\bar{\mu}_1(P) - \bar{\mu}_2(P)| \leq 1$$

and

$$\|\bar{\mu}_1, \bar{\mu}_2\| = \sum_{k=1}^{\infty} \frac{1}{2^k} \|\bar{\mu}_1(P_k) - \bar{\mu}_2(P_k)\|.$$

Lemma 6.2 $\lim_{i \to \infty} \|\bar{\mu}_i, \bar{\mu}\| = 0$ *if and only if for all* $A \in \mathscr{F}$ *and* $B \in \mathscr{G}$,

$$\lim \bar{\mu}_i(A \times B) = \bar{\mu}(A \times B).$$

Thus although the metric depends on the choice of P_i, the topology itself does not.

Proof Certainly if

$$\bar{\mu}_i(A \times B) \underset{i}{\rightarrow} \bar{\mu}(A \times B)$$

then

$$\|\bar{\mu}_i, \bar{\mu}\| \underset{i}{\rightarrow} 0.$$

On the other hand, for any $A \times B$ and $\varepsilon > 0$, select $A' \times B'$ with

$$\mu(A \,\Delta\, A') < \varepsilon \qquad \text{and} \qquad \nu(B \,\Delta\, B') < \varepsilon, \qquad A', B' \in P_i \qquad \text{for some } i.$$

Thus $\hat{\mu}(A \times B \,\Delta\, A' \times B') < 2\varepsilon$ for any $\hat{\mu} \in J(\bar{X}, \bar{Y})$.

$$\lim_{i \to \infty} |\bar{\mu}_i(A' \times B') - \bar{\mu}(A' \times B')| = 0$$

so

$$\lim_{i \to \infty} |\bar{\mu}_i(A \times B) - \bar{\mu}(A \times B)| < 4\varepsilon. \qquad \blacksquare$$

Theorem 6.3 *$(J(\bar{X}, \bar{Y}), \|\cdot, \cdot\|)$ is compact, convex and if $\bar{X}$ and $\bar{Y}$ are ergodic, its extreme points are the ergodic joinings.*

Proof Given any sequence $\bar{\mu}_i \in J(\bar{X}, \bar{Y})$, we can select a subsequence $\bar{\mu}_{i(t)}$ so that $\bar{\mu}_{i(t)}(C)$ is convergent for all sets C in the tree $\{P_i\}$. Let $\bar{\mu}_0(C)$ be the limit. This will be a $T \times S$-invariant additive set function on the tree with marginals μ and ν. Hence $\bar{\mu}_0$ extends to a joining $\bar{\mu}$. Clearly $\|\bar{\mu}_{i(t)}, \bar{\mu}\| \underset{t}{\rightarrow} 0$.

To identify the extreme points, suppose $\bar{\mu} \in J(\bar{X}, \bar{Y})$ and we write $\bar{\mu} = \int_0^1 \bar{\mu}_t \, dt$, its ergodic decomposition. We want to show that for a.e. t, $\bar{\mu}_t \in J(\bar{X}, \bar{Y})$. What we must show is that the marginals of $\bar{\mu}_t$ are μ and ν respectively.

For a set $A \in \mathscr{F}$,

$$\frac{1}{n} \sum_{i=0}^{n-1} \chi_A(T^i(x)) \underset{n}{\rightarrow} \mu(A), \quad \mu\text{–a.e.}$$

Thus $\bar{\mu}$–a.e.,

$$\frac{1}{n} \sum_{i=0}^{n-1} \chi_{A \times Y}(T^i(x), S^i(y)) \underset{n}{\rightarrow} \mu(A).$$

Thus for a.e. t, for all sets A_i, for $\bar{\mu}_t$–a.e. (x, y), this Cesáro limit is $\mu(A_i)$.

Just repeating then, for a.e. t, $\bar{\mu}_t(A_i \times Y) = \mu(A_i)$. The first marginal of $\bar{\mu}_t$ (and by symmetry the second) is μ (is ν). Thus a non-ergodic measure is not extreme. As an ergodic measure cannot be written as an average of any other, an ergodic measure is extreme. ■

6.2 The relatively independent joining

We want to describe now a basic construction. When two dynamical systems have a common factor algebra, this leads to a common supersystem or joining. This is a version in measure-algebraic terms of what is called a fibre product in topology. The construction occurs entirely on the measure space level. That the measure we build on $X \times Y$ is $T \times S$-invariant makes it a joining and what this joining does is to identify the two copies of the common factor algebra, and on the fibres over the factor to take the direct product of the corresponding fibre measures.

To be explicit, let $\bar{X}$ and $\bar{Y}$ be two systems. Suppose $\bar{Z}_1 = (Z_1, \mathscr{H}_1, \eta_1, U_1)$ and $\bar{Z}_2 = (Z_2, \mathscr{H}_2, \eta_2, U_2)$ are factors of these. That is, we have homomorphisms

$$\varphi_1 : \bar{X} \rightarrow \bar{Z}_1$$

$$\varphi_2 : \bar{X} \rightarrow \bar{Z}_2.$$

Further, suppose $\psi : \bar{Z}_1 \rightarrow \bar{Z}_2$ is an isomorphism.

We want to build a joining of $\bar{X}$ and $\bar{Y}$ that identifies sets $\varphi_1^{-1}(C) \in \mathscr{F}$ and $\varphi_2^{-1} \circ \psi(C) \in \mathscr{F}$ as a.s. equal.

We define the joining on rectangles $A \times B$. For $A \in \mathscr{F}$, $E(\chi_A | \varphi_1^{-1}(\mathscr{H}_1))$ is a constant on sets $\varphi_1^{-1}(z_1)$ for η_1–a.e. $z_1 \in Z_1$. Thus

$$f_A(z_1) = E(\chi_B | \varphi_1^{-1}(\mathscr{H}_1)) \circ \varphi_1^{-1}(z_1)$$

is well defined a.e. on Z_1. Similarly

$$g_B(z_2) = E(\chi_B | \varphi_2^{-1}(\mathscr{H}_2)) \circ \varphi_2^{-1}(z_2)$$

is well defined a.e. on Z_2. We have simply projected the conditional expectations down to Z_1 and Z_2 respectively.

Define the set function

$$\begin{aligned}\bar{\mu}_0(A \times B) &= \int_{Z_1} f_A(z_1) \times g_B(\psi(z_1))\,\mathrm{d}\eta_1(z_1) \\ &= \int_{Z_2} f_A(\psi^{-1}(z_2)) \times g_B(z_2)\,\mathrm{d}\eta_2(z_2) \qquad (6.2)\end{aligned}$$

as ψ is measure-preserving.

Lemma 6.4 *The set function $\bar{\mu}_0$ above is additive, $T \times S$-invariant and has μ and v as marginals.*

Proof Additivity follows from linearity of the conditional expectation. Invariance under $T \times S$ follows from

$$E(\chi_{T^{-1}(A)} | \varphi_1^{-1}(\mathscr{H}))(x_1) = E(\chi_A | \varphi_1^{-1}(\mathscr{H}))(T(x_1)).$$

Computing marginals,

$$\bar{\mu}_0(A \times Y) = \int_{A_1} f_A(z_1)\,\mathrm{d}\eta_1 = \int_X E(\chi_A | \varphi_1^{-1}(\mathscr{H}))\,\mathrm{d}\mu = \mu(A). \qquad \blacksquare$$

Thus $\bar{\mu}_0$ extends to a joining $\bar{\mu}$ we call the *relatively independent joining* over $(\bar{Z}_1, \bar{Z}_2, \psi)$.

Lemma 6.5 *For $\bar{\mu}$, the relatively independent joining over $(\bar{Z}_1, \bar{Z}_2, \psi)$,*

(1) $\bar{\mu}(\{(x, y) : \psi(\varphi_1(x)) = \varphi_2(y)\}) = 1$;

(2) $A \in \varphi_1^{-1}(\mathscr{H}_1)$ *if and only if there is a* $B \in \mathscr{G}$ *with*

$$\bar{\mu}(A \times Y \Delta X \times B) = 0;$$

(3) $B \in \varphi_2^{-1}(\mathscr{H}_2)$ *if and only if there is an* $A \in \mathscr{F}$ *with*

$$\bar{\mu}(A \times Y \Delta X \times B) = 0;$$

and

(4) *for such a pair* A *and* B,

$$\psi(A) = B.$$

Proof (1) Let $\{H_i\}$ be a refining and generating tree of partitions in Z_1. Hence $\{\psi(H_i)\}$ is such in Z_2.

Write $H_i = \{A_{i,1}, A_{i,2}, \ldots, A_{i,s(i)}\}$. In the tree $\{\varphi_1^{-1}(H_i) \times \varphi_2^{-1}(\psi(H_i))\}$ on $X \times Y$ are certain special diagonal sets of the form

$$\varphi_1^{-1}(A_{i,j(i)}) \times \varphi_2^{-1}(\psi(A_{i,j(i)})).$$

A chain of such diagonal sets descends to a set of points (x, y) where $\psi(\varphi_1(x)) = \varphi_2(y)$.

Consider

$$\bar{A}_i = \bigcup_{j=1}^{s(i)} (\varphi_1^{-1}(A_{i,j}) \times \varphi_2^{-1}(\psi(A_{i,j}))),$$

the points in diagonal sets at level i. These are nested and $\bar{A} = \bigcap_i \bar{A}_i$ is exactly the points (x, y) with $\psi(\varphi_1(x)) = \varphi_2(y)$.

We compute

$$\begin{aligned}\bar{\mu}(\bar{A}_i) &= \sum_{j=1}^{s(i)} \int_{Z_1} f_{\varphi_1^{-1}(A_{i,j})}(z_1) \times g_{\varphi_2^{-1}(A_{i,j})}(\psi(z_1))\,\mathrm{d}\eta_1(z_1)\\ &= \sum_{j=1}^{s(i)} \int_{Z_1} \chi_{A_{i,j}}(z_1) \times \chi_{\psi(A_{i,j})}(\psi(z_1))\,\mathrm{d}\eta_1(z_1)\\ &= \sum_{j=1}^{s(i)} \eta_1(A_{i,j}) = 1.\end{aligned}$$

Thus $\bar{\mu}(A) = 1$.

(2), (3) and (4): If $A \in \psi_1^{-1}(\mathscr{H}_1)$ then just as above, we compute

$$\bar{\mu}(A \times Y \Delta X \times \psi(A)) = 0.$$

Conversely, suppose

$$0 = \bar{\mu}(A \times Y \Delta X \times B) = \int_{Z_1} f_A(1 - g_B \circ \psi)\,\mathrm{d}\eta_1 + \int (1 - f_A) g_B \circ \psi\,\mathrm{d}\eta_1.$$

Now both f_A and $g_B \circ \psi$ lie between 0 and 1.

For this to be zero, both integrals must be 0. Thus when $g_B \circ \psi(z_1) \neq 1$, we must have $f_A(z_1) = 0$ and when $f_A(z_1) \neq 1$, we must have $g_B \circ \psi(z_1) = 0$ (η_1–a.s. of course). Thus η_1–a.s., if $f_A(z_1) = 0$, then $g_B \circ \psi(z_1) = 0$ and whenever $f_A(z_1) \neq 0$,

$$g_B \circ \psi(z_1) = f_A(z_1) = 1.$$

Thus $f_A = g_B \circ \psi$ is a characteristic function. It follows that $A \in \varphi_1^{-1}(\mathscr{H}_1)$, $B \in \varphi_2^{-1}(\mathscr{H}_2)$ and $B = \psi(A)$. ■

We can rephrase this as the commuting of a certain diagram:

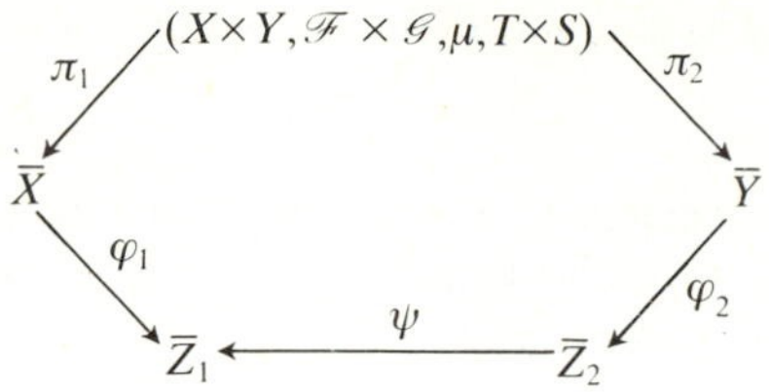

where π_1 and π_2 are the coordinate projections.

Embedded in the relatively independent joining is a single copy of the factor, identified as both $\pi_1^{-1} \circ \varphi_1^{-1}(\mathscr{H}_1)$ and $\pi_2^{-1} \circ \varphi_2^{-1}(\mathscr{H}_2)$.

Lemma 6.6 *For functions $f \in L^1(\mu)$, $g \in L^1(\nu)$,*

$$\begin{aligned} &E_{\bar{\mu}}(f \circ \pi_1 \times g \circ \pi_2 | \pi_1^{-1} \circ \varphi_1(\mathscr{H}_1))(x, y) \\ &\quad = E_{\bar{\mu}}(f \circ \pi_1 \times g \circ \pi_2 | \pi_2^{-1} \circ \varphi_2(\mathscr{H}_2))(x, y) \\ &= E_\mu(f | \varphi_1^{-1}(\mathscr{H}_1))(x) E_\nu(g | \varphi_2^{-1}(\mathscr{H}_2))(y). \end{aligned}$$

Proof (Note: in any expression where the measure under consideration could be ambiguous, we will put the measure as a subscript). We show the identity for characteristic functions.

As $\pi_1^{-1}(\varphi_1^{-1}(\mathscr{H}_1)) = \pi_2^{-1}(\varphi_2^{-1}(\mathscr{H}_2))$ $\bar{\mu}$–a.s.

$$E_\mu(\chi_A | \varphi_1^{-1}(\mathscr{H}_1))(x) E_\nu(\chi_B | \varphi_2^{-1}(\mathscr{H}_2))(y) \tag{6.3}$$

is $\pi_1^{-1}\varphi_1^{-1}(\mathscr{H}_1)$ measurable.

For any set $D \in \mathscr{H}_1$,

$$\begin{aligned} &\int_{\pi_1^{-1}\varphi_1^{-1}(D)} E_\mu(\chi_A | \varphi_1^{-1}(\mathscr{H}_1))(x) E_\nu(\chi_B | \varphi_2^{-1}(\mathscr{H}_2))(y) \, \mathrm{d}\bar{\mu} \\ &\quad = \int_{\pi_1^{-1}\varphi_1^{-1}(D)} f_A(\varphi_1(x)) g_B(\varphi_2(y)) \, \mathrm{d}\bar{\mu}. \end{aligned}$$

Now $\bar{\mu}$–a.s. $\varphi_1(x) = \psi^{-1}\varphi_2(y) = z_1$. Thus our integral is

$$\begin{aligned} \int_D f_A(z_1) g_B(\psi(z_1)) \, \mathrm{d}\eta_1 &= \int \chi_D(z_1) f_A(z_1) g_B(\psi(z_1)) \, \mathrm{d}\eta_1 \\ &= \int f_{A \cap \varphi_1^{-1}(D)}(z_1) g_B(\psi(z_1)) \, \mathrm{d}\eta_1 \\ &= \int_{Z_1} \chi_{A \cap \varphi_1^{-1}(D)}(x) \chi_B(y) \, \mathrm{d}\bar{\mu} \\ &= \int_{\pi_1^{-1}\varphi_1^{-1}(D)} \chi_A(x) \chi_B(y) \, \mathrm{d}\bar{\mu}. \end{aligned}$$

Conditions (6.3) and (6.4) uniquely define the conditional expectation (Exercise 2.12). ∎

Corollary 6.7 *If we decompose $\bar{X}$ over the factor $\bar{Z}_1$ we get fibre measures μ_{1,z_1}, and similarly decomposing $\bar{Y}$ over $\bar{Z}_2$ we get fibre measures μ_{2,z_2}. Decomposing $(X \times Y, \mathcal{F} \times \mathcal{G}, \bar{\mu})$ over the factor*

$$\pi_1^{-1} \circ \varphi_1^{-1}(\mathcal{H}_1) = \pi_2^{-1} \circ \varphi_2^{-1}(\mathcal{H}_2)$$

the fibre measure at (x, y) is

$$\mu_{1,\varphi_1(x)} \times \mu_{2,\varphi_2(y)} = \mu_{1,z_1} \times \mu_{2,\psi(z_2)}.$$

Proof Examining the proof of Theorem 3.17 where the fibre measures are constructed, they are simply the conditional expectations of Lemma 6.6. ∎

Fig. 6.1 indicates how this fibring of the measure space looks. There are two special cases of this construction we should consider.

Example 1 If $\bar{Z}_1$ and $\bar{Z}_2$ are both the trivial process (the identity transformation on one point), then

$$\bar{\mu}(A \times B) = \int_{Z_1} \mu(A)\nu(B)\,d\eta_1 = \mu(A)\nu(B)$$

and $\bar{\mu} = \mu \times \nu$.

Example 2 If $\bar{X}$ and $\bar{Y}$ are isomorphic, ψ the isomorphism, then using $\bar{Z}_1 = \bar{X}$, $\bar{Z}_2 = \bar{Y}$, by Lemma 6.5,

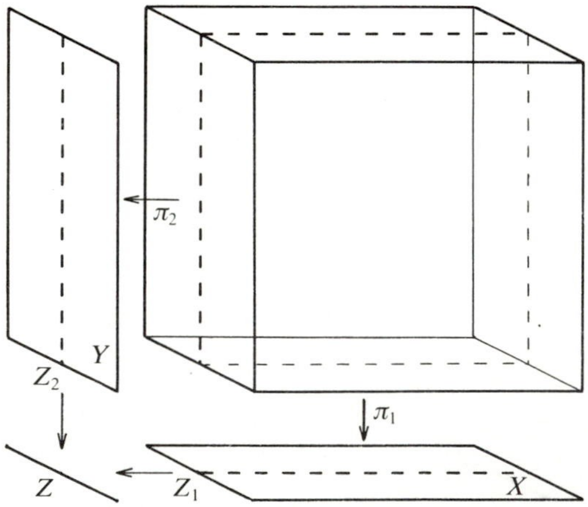

Fig. 6.1 Relatively independent joining over Z. The fibre measure on the dashed square is the direct product of the fibre measures on the two dashed lines.

$$\bar{\mu}(\{(x, \psi(x)) | x \in X\}) = 1.$$

The set of all $(x, \psi(x))$ is just the graph of ψ. This joining is supported on the graph. Thus any automorphism of $\bar{x}$ gives rise to a joining supported on its graph.

This second example has an important converse.

Theorem 6.8 *If $\bar{\mu} \in J(\bar{X}, \bar{Y})$ and $\mathscr{F} \times Y = X \times \mathscr{G}$, $\bar{\mu}$–a.e., then there is an isomorphism $\psi : \bar{X} \to \bar{Y}$ and*

$$\mu(\{(x, \psi(x)) | x \in X\}) = 1.$$

Proof As $\mathscr{F} \times Y = X \times \mathscr{G}$, $\bar{\mu}$–a.s. we can construct refining and generating trees $\{P_i\}$ of X and $\{P_i'\}$ of Y with

$$\bar{\mu}(P_i \times Y \Delta X \times P_i') = 0.$$

To any chain $\mathscr{C}$ in $\{P_i\}$ there corresponds a chain $\mathscr{C}'$ in $\{P_i'\}$. If $\mathscr{C}$ and $\mathscr{C}'$ are not empty, set

$$\psi\left(\bigcap_{c_i \in \mathscr{C}} c_i\right) = \bigcap_{c_i' \in \mathscr{C}'} c_i'.$$

This map takes almost all of X to almost all of Y.

As $\bar{\mu}(A \times Y \Delta X \times \psi(A)) = 0$ for any $A \in P_i$, it is true for all A and ψ is measure-preserving.

As

$$\begin{aligned} v(\psi(T(A)) \Delta S(\psi(A))) &= \bar{\mu}(T(A) \times Y \Delta X \times S(\psi(A))) \\ &= \bar{\mu}(A \times Y \Delta X \times \psi(A)) = 0, \end{aligned}$$

$\psi \circ T = S \circ \psi$, μ–a.s. and ψ is an isomorphism. Following the computation in Lemma 6.5 completes the result. ∎

Theorem 6.8 begins the work of the next chapter, showing us how to identify joinings as isomorphisms.

We now want to apply these ideas. In the remainder of this chapter our applications will involve cases where $J(\bar{X}, \bar{Y})$ is particularly small. In Chapter 7 we will consider applications involving $J(\bar{X}, \bar{Y})$ particularly large.

6.3 Disjointness

Definition 6.3 We say two systems $\bar{X}$ and $\bar{Y}$ are *disjoint* if $J(\bar{X}, \bar{Y}) = \{\mu \times v\}$, i.e., the only joining is product measure.

Exercise 6.1 Show that if α and β are two irrational numbers, and further α/β is irrational, that rotation by α and by β on $\mathbb{R}/\mathbb{Z}$ are disjoing (hint: show $R_\alpha \times R_\beta$ is minimal).

We have already seen that certain mixing properties are characterized by the *non*-existence of certain factors. Thus a map is ergodic if it has no invariant factor, weakly mixing if it has no isometric factor, and K if it has no zero entropy factor. We will now see that something stronger is true, in each case the mixing property implies a disjointness property.

All three of these arguments rest on a seemingly trivial observation. If a sequence $f_i \in L^2(\mu)$ is convergent, then if you embed $L^2(\mu)$ in some larger L^2-space, the f_i will still converge.

Theorem 6.9 *A system $\bar{X}$ is ergodic iff it is disjoint from any identity map on a Lebesgue space.*

Proof The if direction is clear. Suppose T is ergodic, and $\bar{Y} = \{Y, \mathscr{G}, \nu, S\}$ is an identity map. For any $f \in L^2(\mu)$, the ergodic averages $A_n(f)$ converge in $L^2(\mu)$ to the constant $\int f \, d\mu$.

For $\bar{\mu} \in J(\bar{X}, \bar{Y})$, the lift

$$(f \circ \pi_1) \in L^2(\bar{\mu}), \quad A_n(f \circ \pi_1)$$

still converges $L^2(\bar{\mu})$ to $\int f \, d\mu$. Thus for any $g \in L^2(\nu)$,

$$\begin{aligned}\int (f \circ \pi_1) \times (g \circ \pi_2) \, d\bar{\mu} &= \int A_n(f \circ \pi_1 \times g \circ \pi_2) \, d\bar{\mu} \\ &= \int A_n(f \circ \pi_1) \times g \circ \pi_2 \, d\bar{\mu} \quad (\text{as } S = \text{id}) \\ &= \int f \, d\mu \int g \, d\nu.\end{aligned}$$

But this says $\bar{\mu} = \mu \times \nu$. ■

Theorem 6.10 *A system $\bar{X}$ is weakly mixing iff it is disjoint from all isometries of compact metric spaces.*

Proof The if direction is clear. Suppose $\bar{X}$ is weakly mixing and $\bar{Y}$ an isometry of a compact metric space. Following similar lines as the last argument, let $f \in L^2(\mu)$, $g \in L^2(\nu)$ be characteristic functions.

As S is isometric we know (Exercise 4.1) for any $\varepsilon > 0$, there is a sequence $\{n_k\} \subseteq \mathbb{Z}^+$ of positive density with

$$\|g(s^{n_k}(y)) - g(y)\|_2 < \varepsilon.$$

As T is weakly mixing, by Lemma 4.4, there is a sequence $\{m_k\} \subseteq \mathbb{Z}^+$ of full density with $f \circ T^{m_k}$ converging in $L^2(\mu)$ to the constant function $\int f \, d\mu$.

Intersecting these sequences gives an infinite sequence $\{t_k\}$. For $\bar{\mu} \in J(\bar{X}, \bar{Y})$ we compute

$$\left|\int f\circ\pi_1\times g\circ\pi_2\,\mathrm{d}\bar\mu-\int f\,\mathrm{d}\mu\int g\,\mathrm{d}\nu\right|$$
$$=\lim_{k\to\infty}\left|\int f\circ T^{t_k}\circ\pi_1\times g\circ S^{t_k}\circ\pi_2\,\mathrm{d}\bar\mu-\int f\,\mathrm{d}\mu\int g\,\mathrm{d}\nu\right|$$
$$\le\lim_{k\to\infty}\left|\int f\circ T^{t_k}\circ\pi_1\circ g\circ\pi_2\,\mathrm{d}\bar\mu-\int f\,\mathrm{d}\mu\int g\,\mathrm{d}\nu\right|+\varepsilon\|f\|_2$$
$$=\varepsilon\|f\|_2,$$

as $(f\circ T^{t_k}\circ\pi_1)\to\int f\,\mathrm{d}\mu$ in $L^2(\bar\mu)$. Thus $\bar\mu=\mu\times\nu$. ■

Theorem 6.11 *An ergodic $\bar X$ has the K-property iff it is disjoint from all ergodic $\bar Y$ of zero entropy.*

(Note: the ergodicity of $\bar Y$ can be removed.)

Proof The if direction, as always, is already clear. For the other, assume $\bar X$ is K, $\bar Y$ is zero entropy and ergodic, and $\bar\mu\in J(\bar X,\bar Y)$. By Theorem 6.3 we can assume $\bar\mu$ is ergodic. Let P be a finite partition of X and Q of Y. We lift $\bar P=P\times Y$ and $\bar Q=X\times Q$.

From Corollary 5.28, with $\bar\mu$ as measure,

$$D_{\bar\mu}(\bar P|\mathscr{P}_{\bar P}\vee\mathscr{T}_{\bar Q})=D_\mu(P|\mathscr{P}_P)\circ\pi_1.$$

Consider this same identity with $(T\times S)$ replaced by $(T\times S)^{2^k}$ and write it

$$D_{\bar\mu}(\bar P|\mathscr{P}^k_{\bar P}\vee\mathscr{T}^k_{\bar Q})=D_\mu(P|\mathscr{P}^k_P)\circ\pi_1.\tag{6.5}$$

By Lemma 5.6, $h(S^{2^k})=0$, so

$$\mathscr{T}^k_{\bar Q}=\mathscr{T}_{\bar Q}=X\times\bigvee_{i=-\infty}^{\infty}S^{-i}(Q).$$

Definition 4.4 of the K-property says

$$D_\mu(P|\mathscr{P}^k_P)\underset{k}{\to}D(P)=(\mu(p_1),\ldots,\mu(p_s)).$$

Both sides of (6.5) are vector-valued reverse martingales as

$$\mathscr{P}^{k+1}_{\bar P}\subseteq\mathscr{P}^k_{\bar P}.$$

Thus by Theorem 5.30,

$$D_{\bar\mu}\left(\bar P\,\middle|\,\bigcap_k\mathscr{P}^k_{\bar P}\vee\mathscr{T}_{\bar Q}\right)=D(P).$$

Thus

$$\int I\left(\bar{P}\,\middle|\,\bigcap_k \mathscr{P}_{\bar{P}}^k \vee \mathscr{T}_{\bar{Q}}\right)\mathrm{d}\bar{\mu} = \int I(P)\,\mathrm{d}\mu$$

But $\int I(\bar{P}|\bar{Q})\,\mathrm{d}\bar{\mu}$ is squeezed between these two by Corollary 5.26. Thus it also is $\int I(P)\,\mathrm{d}\mu$. But then $\bar{P} \perp \bar{Q}$ relative to $\bar{\mu}$. ∎

We have already pointed out the common thread in the 'only if' direction of these three theorems. In the case of the K-property we were not being quite honest, as the argument was entropy and not L^2-based. Notice that the 'if' directions also are completely parallel. When disjointness from identities, isometries or zero entropy fails it is because of the existence of a factor of this type. We will see in the next seciton that this is not a totally general phenomenon. Systems can fail to be disjoint without possessing common factors.

Each of these results concerns two collections of transformations, characterizing one as the systems disjoint from the other. One can ask if the other member of the pair (identities, isometries, zero entropy) is characterized similarly. For identity maps this is false. There are maps which are disjoint from all ergodic maps but are not the identity. A nice example is $T(x, y) = (x, x + y)$ on the two-dimensional torus. For isometries it is also false. Glasner and Weiss (1990) have constructed systems which are not isometries of compact metric spaces but are disjoint from all weakly mixing systems. For zero entropy though it is true. As part of our work in Chapter 7 (Theorem 7.24) we will see that any ergodic positive entropy system has a Bernoulli shift factor, hence is not disjoint from the K-systems.

6.4 Minimal self-joinings

When one considers two distinct systems it is possible, as we have seen, for $J(\bar{X}, \bar{Y})$ to contain only product measure. When the two systems are the same, $J(\bar{X}, \bar{X})$ must contain more. Elements of $J(\bar{X}, \bar{X})$ we refer to as *self-joinings.*

There are automatically certain automorphisms of this system, the powers of T. The measures supported on their graphs are self-joinings. Thus δ_j supported on the graph $\{(X, T^j(x))\}$ is in $J(\bar{X}, \bar{X})$. We call δ_j an *off-diagonal* as it generalizes diagonal measure δ_0.

Definition 6.4 We say $\bar{X}$ has *two-fold minimal self-joinings* if $J(\bar{X}, \bar{X})$ is the convex hull of $\{\mu \times \mu, \delta_j\}_{j=-\infty}^{\infty}$.

We call this two-fold minimal self-joinings because we could always consider joining any finite number of copies of $\bar{X}$ together. In this case there are again certain joinings that must exist. Define an off-diagonal joining to be one supported on a graph $\{(x, T^{j_1}(x), \ldots, T^{j_k}(x))\}$ in a $(k + 1)$-fold joining. These must be in $J(\bar{X}, \ldots, \bar{X})$. More generally, one could partition the k copies into subsets. On each subset place an off-diagonal measure, then take the direct product of these off-diagonals.

Definition 6.5 We say $\bar{X}$ has *k-fold minimal self-joinings* if all k-fold self-joinings of $\bar{X}$ are in the convex hull of such products of off-diagonals.

In these definitions we are not assuming product measure is necessarily ergodic. We will see though that if a system has minimal self-joinings and is not weakly mixing, then it must be a finite rotation.

We will use this notion as a constructive tool to exhibit control of certain aspects of a dynamical system not directly reachable with our current methods. Our next theorem sets the tone for these ideas.

Theorem 6.12 *If $\bar{X}$ is totally ergodic and has two-fold minimal self-joinings, then it commutes only with its powers and has no non-trivial factor algebras.*

Proof Suppose $T \circ S = S \circ T$, and of course S is measure-preserving. First assume S is invertible. Construct the joining $\bar{\mu}$ supported on the graph of S (Example 6.2).

Now $\bar{\mu}$–a.s. the two coordinate algebras are equal, and $\bar{\mu}$ is ergodic for $T \times T$. Thus $\bar{\mu}$ is either $\mu \times \mu$ or some δ_j. It cannot be $\mu \times \mu$, as this does not identify the coordinate algebras. Thus $\bar{\mu} = \delta_j$ for some j. But then

$$\bar{\mu}(\{(x, S(x))\} \,\Delta\, \{(x, T^k(x))\}) = 1$$

and some $S(x) = T^j(x)$ μ–a.s.

If S is not invertible, then $S^{-1}(\mathscr{F})$ does not separate points and is a non-trivial factor algebra. All that remains, then, is to show there are no such.

Let $\varphi : \bar{X} \to \bar{Z}$ be a factor map. Let $\bar{\mu}$ be the relatively independent self-joining of $\bar{X}$ over $\bar{Z}$. Now $\bar{\mu}$ may not be ergodic, but it must be of the form

$$\mu = \alpha(\mu \times \mu) + (1 - \alpha) \sum_{j=-\infty}^{\infty} a_j \delta_j,$$

where $0 \leq \alpha \leq 1$, and $a_j \geq 0, \sum a_j = 1$. Suppose $\bar{Z}$ is non-trivial, i.e., there is a set $A \in \varphi^{-1}(\mathscr{H})$ with $\mu(A) \neq 0, 1$.

By Lemma 6.5,

$$\mu(A) = \bar{\mu}(A \times A) = \alpha\mu(A)^2 + (1 - \alpha) \sum_{j=-\infty}^{\infty} a_j \mu(A \cap T^{-j}(A)).$$

Of the numbers $\mu(A)^2$ and $\mu(A \cap T^{-j}(A))$ only one is as large as $\mu(A)$, and that is $\mu(A \cap T^{(-0)}(A))$. All the others are strictly smaller, as T is totally ergodic. But then we must have $\alpha = 0$ and $a_j = 0$ unless $j = 0$. We conclude

$$\bar{\mu} = \delta_0,$$

and

$$\bar{\mu}(A \times X) = \bar{\mu}(A \times A) = \bar{\mu}(X \times A)$$

for all $A \in \mathscr{F}$ and so again by Lemma 6.5, $\varphi^{-1}(\mathscr{H}) = \mathscr{F}$ and φ is an isomorphism. ■

Corollary 6.13 (Of the proof). *If $\bar{X}$ has two-fold minimal self-joinings but is not totally ergodic, then X is finite.*

Proof As $\bar{X}$ is not totally ergodic, it has a factor algebra that is a finite point space (Exercise 3.3). Constructing $\bar{\mu}$ the relatively independent self-joining over this factor the above proof still says $\alpha = 0$ as $\mu(A)^2 < \mu(A)$.

But now, for any point $z \in Z$, by Corollary 6.7, the fibre measure of $\bar{\mu}$ over z is $\mu_{1,z} \times \mu_{2,z}$. Hence the fibre measure over the entire first coordinate algebra is $\mu_{2,z}$. But this measure is atomic, equal to $\sum_{j=-\infty}^{\infty} a_j \delta(T^j(x))$ over x. Thus Z is atomic, and the fibre measures μ_z over Z are atomic. Thus X itself is atomic. ■

In fact, any cyclic permutation on a finite set of points has minimal self-joinings. If such were all of them this idea would not be very useful.

6.5 Chacón's map once more

We will show here that Chacón's map has minimal self-joinings of all orders. This argument is due to del Junco, Rahe and Swanson (1980). As always with Chacón's map, it is the spacer placed between the second and third blocks in the cutting and stacking that brings off the argument. By the time we are finished we will have come to a very precise understanding of the name structure of the transformation. The core of the proof is to show that if a pair of points (x, y) gives the expected limit values for the ergodic theorem applied to some self-joining $\bar{\mu}$, but lie on distinct orbits of T, then in fact $\bar{\mu} = \mu \times \mu$. What we have to show is that for such a pair (x, y) the Cesáro averages are those of product measure.

Our first technical lemmas tell us how to recognize product measure.

Lemma 6.14 *If $\bar{X}$ and $\bar{Y}$ are ergodic and $\bar{\mu} \in J(\bar{X} \times \bar{Y})$ is* (id $\times$ S)*-invariant, then $\bar{\mu} = \mu \times \nu$.*

Proof As S is ergodic, by Theorem 6.8, $\bar{\mu} = \mu \times \nu$. ■

Our next lemma tells us how we see (id $\times$ S)-invariance in a $\bar{\mu} \in J(\bar{X}, \bar{X})$. It is far easier to prove than express.

Lemma 6.15 *Suppose $\bar{X}$ and $\bar{Y}$ are ergodic, P is a finite generating partition for $\bar{X}$, and Q for $\bar{Y}$. Let*

$$\bar{\mathscr{A}} = \bigcup_{n=1}^{\infty}\left(\bigvee_{i=-n}^{n}(T \times S)^{-i}(P \times Q)\right),$$

a countable generating algebra of cylinder sets.

Suppose $\bar{\mu} \in J(\bar{X}, \bar{X})$ is ergodic and $(x, y) \in X \times Y$ is a point satisfying

$$\frac{1}{n}\sum_{i=0}^{n-1} \chi_A(T^i(x), S^i(y)) \underset{n}{\rightarrow} \bar{\mu}(A) \tag{6.6}$$

and

$$\frac{1}{n}\sum_{i=0}^{n-1} \chi_A(T^{-i}(x), S^{-i}(y)) \underset{n}{\rightarrow} \bar{\mu}(A) \tag{6.7}$$

for all $A \in \bar{\mathscr{A}}$.

Suppose there are intervals $(i_j, j_k) \subseteq \mathbb{Z}$, $i_k \leq 0 \leq j_k$, $(j_k - i_k) \rightarrow \infty$, an $\alpha > 0$ and subintervals I_k, $I_k + t_k \subseteq (i_k, j_k)$ with $\#(I_k) \geq \alpha(j_k - i_k + 1)$.

Finally, for $i \in I_k$,

(1) $P(T^i(x)) = P(T^{i+t_k}(x))$; *and*

(2) $Q(S^i(y)) = Q(S^{i+t_k+1}(y))$. (6.8)

We conclude $\bar{\mu}$ is id $\times$ *S-invariant and hence $\bar{\mu} = \mu \times \mu$.*

Proof Let $(C \times C') \in \bigvee_{i=-n_0}^{n_0}(T \times S)^{-i}(P \times Q)$ be some cylinder set. We want to show $\bar{\mu}(C \times S(C')) = \bar{\mu}(C \times C')$. This of course completes the result.

We have convergence of the Birkhoff theorem in both directions for $T \times S$ on cylinder sets. The intervals I_k and $I_k + t_k$ occupy a fraction $\alpha > 0$ of (i_k, j_k). It is a simple argument that

$$\lim_{k\rightarrow\infty} \frac{1}{\#(I_k)} \sum_{i \in I_k} \chi_{C \times C'}(T^i(x), S^i(y)) = \bar{\mu}(C \times C') \tag{6.9}$$

and

$$\lim_{k\rightarrow\infty} \frac{1}{\#(I_k)} \sum_{i \in I_k + t_k} \chi_{C \times S(C')}(T^i(x), S^i(y)) = \bar{\mu}(C \times S(C')). \tag{6.10}$$

(Break (i_k, j_k) into pieces and consider those on which this convergence *must* occur. This forces convergence on any differences of such pieces that occupy at least a fraction $\alpha > 0$ of (i_k, j_k).)

But now for i more than $2n_0 + 1$ positions from the ends of I_k, (6.8) tells us $(T \times S)^i(x, y) \in C \times C'$ exactly when $(T \times S)^{i+t_k}(x, y) \in C \times S(C')$. But then by (6.9) and (6.10)

$$\left|\frac{1}{\# I_k}\left(\sum_{i \in I_k} \chi_{C \times C'}(T^i(x), S^i(y)) - \sum_{i \in I_k + t_k} \chi_{C \times S(C')}(T^i(x), S^i(y))\right)\right| \leq \frac{4n_0 + 2}{\# I_k}.$$

Letting $k \rightarrow \infty$ we are done. ■

What we need to do now is to show that we can find the structure of this lemma in Chacón's map. We first exhibit a generating partition.

Lemma 6.16 *Let $\bar{X}$ be Chacón's map. The two-set partition P of X into points in the zero block, and its complementary set of points added in spacers, generate.*

Proof We argue inductively that the T,P-name of a point determines its level in a tower. This is equivalent to saying the T,P-name can be broken in only one way into n-blocks, each of which is the name of a passage up through the n-tower.

This is true for 0-blocks, as those are the individual occurrences in the name of p_1.

Suppose we can recognize passages through the n-tower uniquely in the T,P-name of x. These blocks are disjoint and separated in the name by at most one occurrence of p_2. Such single p_2's indicate a point on the orbit of x added as a spacer after stage n.

Whenever we see an n-block with such a spacer both above and below it, this n-block must be the third (top) n-block in an $(n+1)$-block. We must see such an n-block at least one out of every nine. Once we see one such the entire name breaks uniquely into triples of n-blocks forming $(n+1)$-blocks. ■

This argument has begun an investigation of T,P-names in $\bar{X}$, telling us such a name breaks up into a hierarchy of n-blocks, the n-blocks grouping into triples to form the $(n+1)$-blocks.

Let $N(n)$ be the number of levels in the nth stack, hence the length of a passage through an n-block. List these intervals as $I(1,n), \ldots, I(N(n),n)$ from bottom to top. Thus $T(I(i,n)) = I(i+1,n)$ for $i < N(n)$. Let $h_n(x)$ to be the index with$x \in I(h_n(x)+1,n)$. It is undefined if x is not in the nth stack.

Set $k_n(x)$ to be 1, 2 or 3 depending on whether x is in the first, second, or third occurrence of an $(n-1)$-stack in the n-stack. Again k_n is undefined if x is not in the $(n-1)$-stack.

In terms of the T,P-name of x, the origin is at a position $h_n(x)$ to the right of the beginning of its n-block. When $(n-1)$-blocks get grouped to form n-blocks, $k_n(x)$ tells us which of the three contain the origin.

Lemma 6.17 *For μ*–a.e. *x, the set of $n \in \mathbb{N}$ with $k_n(x) = 1$, 2 or 3 each has density* 1/3.

Proof For a.e. x, $k_{n_0}(x)$ is defined once n_0 is large enough. Let $\mathscr{T}_{n_0} = \bigcup_{i=1}^{N(n_0)} I(i,n_0)$ be the n_0-tower.

It is a simple induction that the functions $k_{n_0+1}, k_{n_0+2}, \ldots$ restricted to $\mathscr{T}_{n_0}$ are independent and take on values 1, 2, 3 equally likely. By the law of large numbers (just the ergodic theorem applied to the Bernoulli shift (1/3, 1/3, 1/3)), for μ–a.e. $x \in \mathscr{T}_{n_0}$, $k_n(x) = 1$, 2, 3, each with density 1/3 in $\mathbb{N}$. ■

The sequence of functions $k_n(x)$ 'drive' the construction of T. If we know their values we can construct the T,P-name of X, the first index n_0 for which $k_{n_0}(x)$ is defined tells us x is in the spacer of the n_0-block, i.e., $h_{n_0}(x) = 2N(n_0 - 1) + 1$. Each successive value $k_n(x)$ tells us how to extend the name across the n-block. If these extensions covered all of $\mathbb{Z}$ we would get the full name this way. As $k_n(x) = 2$ infinitely often, with probability one we do cover all of $\mathbb{Z}$.

In fact the only points where this procedure does not generate the full name is when either $k_n(x)$ is asymptotically always 1 or always 3. In this case we only get a half name. These two half names can be combined to form a full name, two in fact, one with a spacer between the two half names.

Lemma 6.18 *For μ–a.e. x, a point $y = T^j(x)$ iff $k_n(x) = k_n(y)$ for all large enough n.*

Proof Suppose $y = T^j(x)$. Then once $h_n(x)$ and $N(n) - h_n(x)$ are at least $|j|$, $k_{n+1}(x) = k_{n+1}(y)$. On the other hand, if $k_n(x) = k_n(y)$ for all $n \geq n_0$, then for some $|j| < N(n_0)$, y and $T^j(x)$ have identical T,P-names. As P generates, $y = T^j(x)$. ∎

Theorem 6.19 (del Junco–Rahe–Swanson) *Chacón's transformation has two-fold minimal self-joinings.*

Proof Suppose $\bar{\mu}$ is an ergodic self-joining of $\bar{X}$. Assuming $\bar{\mu} \neq \delta_j$ for any j, $\bar{\mu}$ gives measure 0 to the union of the graphs of all powers of T. Thus for $\bar{\mu}$–a.e. (x, y), $k_n(x) \neq k_n(y)$ for infinitely many values n.

Select (x, y) so that the Birkhoff theorem is satisfied for both $(T \times T)$ and $(T^{-1} \times T^{-1})$ with respect to $\bar{\mu}$ for all sets in

$$\bar{\mathscr{A}} = \bigcup_{n=1}^{\infty} \left(\bigvee_{i=-n}^{n} (T \times T)^{-i}(P \times P) \right).$$

We are going to establish the structure of Lemma 6.14 on the $(T \times T)$, $P \times P$-name of (x, y).

Let $n_1, n_2, \ldots$ be those values with $k_{n_s}(x) \neq k_{n_s}(y)$ and both are defined. We know $\bar{\mu}$–a.s. both $k_n(x)$ and $k_n(y)$ are infinitely often 2, although these may never be at a value n_s.

But one of the two following cases must occur:

(1) For infinitely many values s, either $k_{n_s-1}(x) = 2$; or

(2) for infinitely many $k_{n_s-1}(x) = k_{n_s-1}(y)$. (6.11)

The only way (2) can fail is if $k_n(x) \neq k_n(y)$ for all sufficiently large n. In this case $k_{n_s-1}(x) = 2$, infinitely often.

In either case, in the T,P-name of x and y, the indices of the $(n_s - 1)$-blocks

Table 6.1

$k_{n_s}(x)$ / $k_{n_s}(y)$	T,P-name of x / T,P-name of y
1	$B_0 \; B_1 \; {*}B_2$
2	$B'_{-1} \; B'_0{*}B'_1$
1	$B_0 \; B_1{*}B_2$
3	$B'_{-2}B'_{-1}{*}B'_0(*) \; B'_1 \; B'_2 \; {*}B'_3$
2	$B_{-1} \; B_0{*}B_1$
1	$B'_0 \; B'_1 \; {*}B'_2$
2	$B_{-1} \; B_0 \; {*}B_1$
3	$B'_{-1} \; B'_{-1}{*}B'_0$
3	$B_{-2}B_{-1}{*}B_0(*) \; B_1 \; B_2 \; {*}B_3$
1	$B'_0 \; B'_1{*}B'_2$
3	$B_{-2} \; B_{-1}{*}B_0$
2	$B'_{-1} \; B'_0 \; {*}B'_1$

containing x and y respectively intersect (overlap) in at least $N(n_s - 2) = (N(n_s - 1) - 1)/3$ places.

Consider the partition of the T,P-names of x and y into $(n_s - 1)$-blocks with possible spaces in between. We want to list the various possibilities we might see near the origin depending on the values $k_{n_s}(x)$ and $k_{n_s}(y)$.

Table 6.1 lists them. A B_i indicates an $(n_s - 1)$-block in the T,P-name of x, B'_i in that of y. A * indicates a spaces, (*) a *possible* spacer. The origin lies in B_0 and B'_0.

In all six cases we have identified a value $|t| \leq 1$ where one sees either

$$(1) \quad \left| \frac{B_t^{*} B_{t+1}}{B'_t B'_{t+1}} \right|, \quad \text{or}$$

$$(2) \quad \left| \frac{B_t B_{t+1}}{B'^{*}_t B'_{t+1}} \right|. \tag{6.12}$$

The overlaps $(B_t \cap B'_t)$ and $(B_{t+1} \cap B'_{t+1})$ are each at least $(N(n_s - 1) - 1)/3 - 2$ and as $|t + 1| \leq 2$, are contained in an interval

$$(i_s, j_s) = (-2N(n_s - 1) - 2, 2N(n_s - 1) + 2).$$

In (1) when the spacer is between B_t and B_{t+1}, set $I_s = B_t \cap B'_t$ and $I'_s = B_{t+1} \cap (B'_{t+1} + 1)$.

In (2), set $I_s = B_{t+1} \cap B'_{t+1}$ and $I'_s = B_t \cap (B'_t + 1)$.
In either case

$$\#(I_s) = \#(I'_s) \geq \frac{1}{18}\left(\frac{N(n_s - 1) - 7}{N(n_s - 1) + (4/3)}\right)(j_s - i_s - 1)$$

$$\geq \frac{1}{20}(j_s - i_s + 1),$$

once s is large enough.

The T,P-name across all $(n_s - 1)$-blocks is the same, so condition (6.8) of Lemma 6.15 is established and $\bar{\mu} = \mu \times \mu$. ∎

Theorem 6.20 (del Junco–Rahe–Swanson) *Chacón's transformation has minimal self-joinings of all orders.*

Proof We know the result for $k = 2$. Assume it for some value $k \geq 2$. An off-diagonal joining of any number of copies of $\bar{X}$ is isomorphic to a single copy. Thus any $(k + 1)$-fold joining which, when restricted to some pair of copies, is an off-diagonal is in fact a joining of just k copies. By inducion, then, we can assume that on any subset of k of the copies of $\bar{X}$ we have product measure.

What we will do is show that $\bar{\mu}$ is either $(\text{id})^{(k-1)} \times T \times \text{id}$- or $(\text{id})^{k-1} \times (T \times T)$-invariant. In either case, Lemma 6.14 completes the result.

In the first k coordinates consider the set S_n of points $(x_1, x_2, \ldots, x_k)$ for which

$$h_n(x_i) \leq \frac{N(n)}{10} \quad \text{for } i = 1, \ldots, k,$$

$$k_{n+1}(x_i) = 1 \quad \text{for } i = 1, \ldots, k - 1,$$

but

$$k_{n+1}(x_k) = 2.$$

As $\bar{\mu}$ is product measure on these k coordinates,

$$\bar{\mu}(S_n) \geq \left[\left(\frac{1}{10} - \frac{1}{N(n)}\right)\frac{1}{3}\right]^k \geq \left(\frac{1}{60}\right)^k \tag{6.13}$$

once $n \geq 3$.

Table 6.2

x_1		$B_1^1 B_1^1$	$*B_2^1$
$\vdots$		$\vdots$	
x_{k-1}		$B_0^{k-1} B_1^{k-1}$	$*B_2^{k-1}$
x_k	B_{-1}^k	$B_0^k * B_1^k$	

Thus $\bar{\mu}(\limsup(S_n)) \geq 0$. As $\bar{\mu}(S_n \Delta (T \times \cdots \times T)S_n) \leq k/N(n) < k/3^n$, $\limsup(S_n)$ is $(T \times \cdots \times T)$-invariant $\bar{\mu}$–a.s. As we assume $\bar{\mu}$ an ergodic joining, $\bar{\mu}$–a.s., $(x_1, \ldots, x_k, x_{k+1}) \in S_n$ for infinitely many n.

Suppose $(x_1, \ldots, x_k, x_{k+1}) \in S_n$ and consider the n-blocks near the origin (Table 6.2).
We know $\bigcap_{i=1}^k B_0^i$ and $\bigcap_{i=1}^k B_1^i$ both have length not less than $9N(n)/10 - 1$.

What do we see in the T,P-name of x_{k+1} across this section? We see two blocks $B_0^{k+1}B_1^{k+1}$ or $B_0^{k+1}*B_1^{k+1}$ where $\bigcap_{i=1}^{k+1} B_0^i$ and $\bigcap_{i=1}^{k+1} B_1^i$ are both $\geq \frac{1}{2}(9N(n)/10 - 1)$. Here B_0^{k+1} is chosen to be the block whose intersection with $\bigcap_{i=1}^k B_0^i$ is largest, not necessarily the one containing the origin.

If for infinitely many n we see $B_0^{k+1}B_1^{k+1}$ then we will conclude, as in Theorem 6.19, that $\bar{\mu}$ is $(\mathrm{id})^{(k-1)} \times T \times \mathrm{id}$-invariant. If $B_0^{k+1}*B_1^{k+1}$ occurs infinitely often, $\bar{\mu}$ is $(\mathrm{id})^{(k-1)} \times (T \times T)$-invariant. In either case, induction forces $\bar{\mu} = (\mu)^k$. ∎

The proofs of Theorems 6.19 and 6.20 followed from a rather simple observation about the name structure of (T, P). The map T^{-1} has this same structure, except the spacer is always placed between the first and second, rather than second and third blocks.

In fact, at each stage of the construction one could make an independent decision as to which one of these two choices we make. Explicitly, let $\mathbf{e} = \{e_1, e_2, \ldots\}$ be an infinite sequence of 0's and 1's. For each such we build a transformation $T_{\mathbf{e}}$ analogous to Chacón's. The spacer at stage n of the construction is placed between the first and second $(n-1)$-blocks if $e_n = 0$, and the second and third if $e_n = 1$. Chacón's map corresponds to $\mathbf{e} = \{1\}$ and its inverse $\mathbf{e} = \{0\}$. In fact, interchanging 1's and 0's in $\mathbf{e}$ always takes $T_{\mathbf{e}}$ to $T_{\mathbf{e}}^{-1}$.

Exercise 6.2

1. Show that all $T_{\mathbf{e}}$ have minimal self-joinings of all orders.
2. Show that if $e_n = e'_n$ for all $n \geq N_0$, then $T_{\mathbf{e}} \cong T_{\mathbf{e}'}$.
3. Show that if $\mathbf{e}$ and $\mathbf{e}'$ differ in infinitely many places, then $T_{\mathbf{e}}$ and $T_{\mathbf{e}'}$ are disjoint. Hint: show any ergodic joining is $\mathrm{id} \times T_{\mathbf{e}'}$-invariant.

Thus Chacón's map is not isomorphic to its inverse.

Exercise 6.3 Show that a map with two-fold minimal self-joinings must have zero entropy.

Exercise 6.4 Show that any non-atomic system with two-fold minimal self-joinings must be weakly mixing.

Thus all the maps $T_{\mathbf{e}}$ are entropy zero, weakly mixing, and if not isomorphic, then disjoint. Notice how much more precise the information gained from the explicit name structure of a system is than the more global mixing or entropy data.

The disjointness observed in Exercise 6.2 parts 2 and 3 is a general fact. Among systems with two-fold minimal self-joinings, two are either disjoint or isomorphic. This result is just the beginning of a much deeper analysis of minimal self-joinings on which we will not embark. We refer the interested reader to bibliographic items del Junco and Keane (1985); del Junco and Rudolph (1987); and Rudolph (1979).

6.6 Constructions

Systems with minimal self-joinings have no factors and commute only with their powers. This is perhaps too simple a situation. From it, though, we can build up examples whose behavior is controlled, but is not quite this trivial.

Start with $\bar{X}$, a non-atomic system with minimal self-joinings of all orders. Let $\bar{X}^k$ be its k-fold direct product. On this we can define a group of transformations as follows. For $i \in \mathbb{Z}$ and π a permutation of $\{1, \ldots, k\}$, set

$$U_{(i,\pi)}(x_1, \ldots, x_k) = (T^i(x_{\pi^{-1}(1)}), \ldots, T^i(x_{\pi^{-1}(k)})). \tag{6.14}$$

Thus $U_{(i,\pi)}$ acts by permuting the k coordinates by π and acting by T^i on each. Notice

$$U_{(i,\pi)} \circ U_{(j,\pi')} = U_{(i+j,\pi\circ\pi')}.$$

We write $\bar{X}_{(i,\pi)}$ to indicate $U_{(i,\pi)}$ acting on the k-fold direct product.

Lemma 6.21 *For any $i \neq 0$ and $\pi, \pi' \in S(k)$ a joining of $\bar{X}_{(i,\pi)}$ and $\bar{X}_{(i,\pi')}$ is in fact a self-joining of $\bar{X}_{(1,\mathrm{id})}$, hence a $2k$-fold self-joining of $\bar{X}$.*

Proof Suppose $\bar{\mu} \in J(\bar{X}_{(i,\pi)}, \bar{X}_{(i,\pi')})$. As $U^{k!}_{(i,\pi)} = U^{k!}_{(i,\pi')} = U_{(j,\mathrm{id})}$, $j = k!i \neq 0$, $\bar{\mu}$ is a self-joining of $U_{(j,\mathrm{id})}$. As we can decompose $\bar{\mu}$ into ergodic components for $U_{(j,\mathrm{id})} \times U_{(j,\mathrm{id})}$, we assume the action is ergodic. Setting

$$\hat{\mu} = \frac{1}{j}\sum_{t=0}^{j-1} \bar{\mu} \circ (U_{(t,\mathrm{id})} \times U_{(t,\mathrm{id})}), \tag{6.15}$$

$\hat{\mu}$ is an ergodic self-joining of $\bar{X}_{(1,\mathrm{id})}$, hence a $2k$-fold self-joining of $\bar{X}$.

Thus $\hat{\mu}$ must be a product of off-diagonal measures. As $\bar{X}$ is weakly mixing, $U_{(j,\mathrm{id})} \times U_{(j,\mathrm{id})}$ acts ergodically on such a $\hat{\mu}$. But (6.14) writes $\hat{\mu}$ as an average of $U_{(j,\mathrm{id})} \times U_{(j,\mathrm{id})}$-invariant measures. Thus they must all the equal to $\hat{\mu}$, i.e., $\bar{\mu} = \hat{\mu}$. The ergodic self-joinings of $\bar{X}_{(j,\mathrm{id})}$ agree with those of $\bar{X}_{(1,\mathrm{id})}$. These are the extreme points of all self-joinings, so

$$J(\bar{X}_{(1,\mathrm{id})}, \bar{X}_{(1,\mathrm{id})}) = J(\bar{X}_{(j,\mathrm{id})}, \bar{X}_{(j,\mathrm{id})}).$$

But $J(\bar{X}_{(i,\pi)}, \bar{X}_{(i,\pi')})$ is squeezed between these. ∎

Example 1 We've already seen that a non-atomic system with two-fold minimal self-joinings could have no roots (a power cannot be a root). Here is an example of a system with more than one non-isomorphic square root. Let $k = 2$, $T_1 = U(i, \mathrm{id}) = T \times T$ and $T_2 = U(1, (1,2))$. We use cycle form to represent π. Thus $T_1^2 = T_2^2 = T^2 \times T^2$. We want to show T_1 and T_2 cannot be isomorphic.

Suppose $\bar{\mu}$ is a joining supported on the graph of an isomorphism. This, by Lemma 6.21, must be a four-fold self-joining of $\bar{X}$. It is product measure on both the first two and second two coordinates, but identifies these two product algebras. Call them $\mathscr{F}_1 \times \mathscr{F}_2$ and $\mathscr{F}_3 \times \mathscr{F}_4$.

As $\bar{\mu}$ is a product of off-diagonals, it must identify $\mathscr{F}_1$ with $\mathscr{F}_3$ or $\mathscr{F}_4$. Suppose $\mathscr{F}_3$. Then

$$\mathscr{F}_3 = \mathscr{F}_1 = (T_1 \times T_2)(\mathscr{F}_1) = (T_3 \times T_4)(\mathscr{F}_3) = \mathscr{F}_4,$$

$\bar{\mu}$–a.s. This is of course false. The same holds if $\mathscr{F}_1$ is identified with $\mathscr{F}_4$. We conclude no isomorphism exists.

Notice that even though T_1 and T_2 are not isomorphic, they do have a common factor, the factor of symmetric sets invariant under interchanging the two coordinates. This factor algebra has two point fibres in both systems. The relatively independent joining over this common factor is ergodic for $T_1 \times T_2$ but not for $T \times T \times T \times T$. It is obtained by averaging two four-fold joinings, one supported on points $\{(x_1, x_2, x_1, x_2)\}$, the other on $\{(x_1, x_2, x_2, x_1)\}$.

Also notice that $T^2 \times T^2$ has other square roots. For example

$$T_3 : (x_1, x_2) \to (x_2, T^2(x_1)).$$

This, though, is isomorphic to T_2 by the isomorphism

$$\varphi : (x_1, x_2) \to (T^{-1}(x_1), x_2).$$

In fact, any square root of $T^2 \times T^2$ is either T_1 or isomorphic to T_2.

Exercise 6.5 Construct a system with at least countably infinitely many pairwise non-isomorphic square roots.

Example 2 It is quite simple to find a system with no non-trivial factors but which commutes with more than its powers. Just take T^2 where T has two-fold minimal self-joinings. To get a system which commutes only with its powers, but has non-trivial factors, we do the following. Let $U = T \times T \times T$. We will take a certain factor of this, the σ-algebra of sets invariant under

$$\rho : (x_1, x_2, x_3) \to (x_2, x_3, x_1).$$

This consists of sets of the form

$$A \cup \rho(A) \cup \rho^2(A), \quad A \in \mathscr{F}_1 \times \mathscr{F}_2 \times \mathscr{F}_3.$$

Let $\bar{Y}$ be U restricted to this factor. Now $\bar{Y}$ has factors. For example those sets also invariant under

$$\rho' : (x_1, x_2, x_3) \to (x_2, x_1, x_3)$$

form a non-trivial subalgebra.

Now $\bar{Y}$ is a factor of $\bar{X} \times \bar{X} \times \bar{X}$ with three point fibres. The fibre consists of the three points (x_1, x_2, x_3), $\rho((x_1, x_2, x_3))$ and $\rho^2((x_1, x_2, x_3))$ which $\bar{Y}$ cannot separate. If $\bar{\mu}$ is a self-joining of $\bar{Y}$, then $\bar{\mu}$ can be extended via the relatively independent joining, to a self-joining of $(T \times T \times T)$.

Suppose φ commutes with S. Support $\bar{\mu}$ on the graph of φ and extend it to $\hat{\mu}$, a self-joining of $(T \times T \times T)$. Now $\hat{\mu}$ still has $\bar{Y}$ as a factor, but has nine point fibres over it, consisting of a choice for x_1 in the first copy, and for x_1' in the second, from among the three. Thus $\hat{\mu}$ is invariant under the nine-element group H generated by $\rho \times \mathrm{id}$ and $\mathrm{id} \times \rho$. The copy of $\bar{Y}$ consists of those sets invariant under this group H.

Now $\hat{\mu}$ may not (in fact, cannot) be an ergodic joining. H permutes the ergodic components of $\hat{\mu}$. As the fibres over $\bar{Y}$ are atomic, in any ergodic component $\hat{\mu}_z$ of $\hat{\mu}$, the coordinate algebras $\mathscr{F}_1$, $\mathscr{F}_2$ and $\mathscr{F}_3$ must be identified bijectively with $\mathscr{F}_4, \mathscr{F}_5, \mathscr{F}_6$. If any one were independent, the fibres would not be atomic. If $\mathscr{F}_1$ with $\mathscr{F}_4$, then automatically $\mathscr{F}_2$ with $\mathscr{F}_5$ and $\mathscr{F}_3$ with $\mathscr{F}_6$. This is an ergodic joining. Acting on it by $\rho \times \mathrm{id}$ and $\rho^2 \times \mathrm{id}$ we get two others, and $\hat{\mu}$ must be an average of such triples.

In μ_z the identification of $\mathscr{F}_1$ with $\mathscr{F}_4$ is via some power of T, i.e., the measure restricted to this pair of algebras is supported on a graph $\{(x_1, T^j(x_1))\}$. Acting by $\rho \times \rho$, there must also be a joining supported on $\{(x_2, T^j(x_2))\}$ and similarly $\{(x_3, T^j(x_3))\}$.

This says that the set of sextuples $\{(x_1, x_2, x_3, T^j(x_1), T^j(x_2), T^j(x_3))\}$ has $\hat{\mu}$-positive measure and as it is $\bar{Y}$ measurable and $T \times T \times T$-invariant, measure 1. We conclude that $\bar{\mu}$ is supported on the graph of S^j and hence this is φ.

In both these examples we considered algebras of sets invariant under some group of coordinate permutations. In our last two examples they will again carry the day.

Exercise 6.6 Let $H \subseteq S(n)$ be a subgroup of the symmetric group. In $(\bar{X})^{(n)}$ consider the algebra $\mathscr{A}_H$ of subsets invariant under the coordinate permutations in H, i.e., under all

$$\rho_h(x_1, \ldots, x_n) = (x_{h^{-1}(1)}, \ldots, x_{h^{-1}(n)}).$$

Let $\bar{Y}_H$ be $(\bar{X})^{(n)}$ restricted to this invariant σ-algebra.

Show that $\bar{Y}_H$ commutes only with its powers iff H acts transitively on $\{1, \ldots, n\}$.

In fact the centralizer and factor algebras of such $\bar{Y}_H$ can be completely characterized in terms of the action of H. We refer the reader to Rudolph (1979) for this discussion.

Example 3 Let $\bar{Y}$ be the factor of symmetric sets in $\bar{X} \times \bar{X}$. Certainly $\bar{Y}$ commutes only with its powers by Exercise 6.6. We want to show it has no non-trivial factors. This follows much the lines of the proof that $\bar{X}$ has no factors. Any such factor is also a factor of $\bar{X} \times \bar{X}$. Let $\bar{\mu}$ be the relatively independent joining over this factor algebra.

Write $\bar{\mu} = \sum a_z \bar{\mu}_z$, its ergodic decomposition, as products of off-diagonals. Suppose the factor contains a non-trivial set A. As $\mu \times \mu(A) = \bar{\mu}(A)$, and $\bar{\mu}_z(A \times A) \leq \mu \times \mu(A)$ for all z, the only possible $\bar{\mu}_z$ are those giving equality. These are supported on one of two graphs:

(1) $\{(x_1, x_2, x_1, x_2)\}$; or

(2) $\{(x_1, x_2, x_2, x_1)\}$.

Both of these, when restricted to $\bar{Y}$, give diagonal measure, and so the factor must be all of $\mathscr{G}$. Notice that $\bar{Y}$ does not have minimal self-joinings, as it can be joined non-trivially in $\bar{X} \times \bar{X} \times \bar{X}$.

Example 4 Consider the two systems $\bar{Y}$ from Example 3, and $\bar{X}$ itself. These two have no common factors, as they each separately have no non-trivial factors, and as only one has minimal self-joinings, they are not isomorphic.

On the other hand, they are joined non-trivially in $\bar{X} \times \bar{X}$, hence are not disjoint.

The reader interested in pursuing the construction of examples like the above, is referred to the bibliography. Some of this development can be subsumed into general classes of maps like our $U(i, \pi)$. Much, though, remains a matter of individual construction and analyses of the structure of names, especially for positive entropy examples.

7 The Krieger and Ornstein theorems

Both Krieger's finite generator theorem (Krieger 1970) and Ornstein's isomorphism theorem (Ornstein 1974) for Bernoulli shifts are representation theorems. Krieger's theorem tells us that any ergodic process of entropy less than $\log_2(n_0)$ can be represented as the shift map on the space $\{p_1, p_2, \ldots, p_{n_0}\}^{\mathbb{Z}}$ with an appropriate shift-invariant Borel measure. Ornstein's theorem characterizes those systems which can be represented again as the shift map on $\{p_1, \ldots, p_{n_0}\}^{\mathbb{Z}}$, only now with a Bernoulli measure.

In both these cases the representing systems are shift maps on a symbol space $\{p_1, \ldots, p_{n_0}\}^{\mathbb{Z}}$ with which is associated a shift-invariant Borel measure. Both theorems are proved by identifying a certain weak* closed collection of joinings of our target system and such symbolic systems. The isomorphisms of the theorems will in fact be dense G_δ-sets in these joining spaces. We will obtain this by identifying a notion of approximate representation and showing that such approximate representations are open and dense. This approach originates in unpublished work of R. Burton and A. Rothstein.

7.1 Symbolic spaces and processes

Let $P = \{p_1, \ldots, p_{n_0}\}$ be a finite state space. Define

$$Y_P = \{p_1, \ldots, p_{n_0}\}^{\mathbb{Z}}, \tag{7.1}$$

the full P-shift. This is a compact metric space in the product topology.

A point $y \in Y_P$ is of the form $(\ldots, y(-n), \ldots, y(0), \ldots, y(n), \ldots)$, an infinite sequence of elements of P. We define the left shift

$$S(y)(n) = y(n+1),$$

a homeomorphism of Y_P. We will always use S for the left shift no matter what the state space P.

There is a natural partition of Y_P labeled by P,

$$P(y) = y(0),$$

and in fact,

$$y = (\ldots, P(S^{-n}(y)), \ldots, P(y), \ldots, P(S^n(y)), \ldots).$$

On Y_P there are many S-invariant measures. We can see this in two ways. First, the generating tree of partitions

$$P_n = \bigvee_{i=-n}^{n} S^{-i}(P) \tag{7.2}$$

has no empty chains. Any finitely additive, S-invariant set function on the tree will extend to such a measure. Notice S-invariance is visible in the tree itself as

$$P_{n+1} = S(P_n) \vee S^{-1}(P_n). \tag{7.3}$$

Thus finite additivity and S-invariance are just a list of identities that a set function on the tree should satisfy.

Let η_P be the space of all S-invariant Borel measures on $(Y_P, \mathscr{B})$. We can explicitly metrize it by

$$\|v_1, v_2\| = \sum_{n=1}^{\infty} \frac{\|v_1(P_n), v_2(P_n)\|}{2^n} \tag{7.4}$$

where $\|v_1(P_n), v_2(P_n)\| = \frac{1}{2}\sum_{\mathbf{p} \in P_n} |v_1(\mathbf{p}) - v_2(\mathbf{p})| \leq 1$, analogous to Definition 6.2. This metric depends only on v_1, v_2 as additive set functions on the tree. The identities which imply v is additive and S-invariant are closed conditions in this metric. Hence $(\eta_{P'} \|\cdot, \cdot\|)$ is a compact space.

It is important to note here that $v_i \to v$ iff $v_i(A) \to v(A)$ for all sets A in the tree, and not for a larger collection of sets, as in Lemma 6.2. This topology on η_P is precisely the classical weak* topology on η as a subset of the dual of the continuous functions. To see this, just notice that for A in the tree, χ_A is continuous. Finite linear combinations of such characteristic functions are uniformly dense in $C(Y_P)$.

Since an average of S-invariant borel probability measures on η_P is again such, η_P is also convex. The taking of convex combinations $\alpha v_1 + (1 - \alpha)v_2$ is jointly continuous in α, v_1 and v_2. The extreme points of η_P we know, by Corollary 3.18, are the ergodic measures. Surprisingly, we will see later the ergodic measures are also dense.

Here is another way to see how rich η_P is. Let $\bar{X} = (X, \mathscr{F}, \mu, T)$ be any system, and $\bar{P}$ any partition of X labeled by P. To the pair $(\bar{X}, \bar{P})$ we can associate an element $v = v_{(\bar{X},\bar{P})} \in \eta_P$. The map

$$\mathbf{p}(x) = \ldots, \bar{P}(T^{-n}(x)), \ldots, \bar{P}(x), \ldots, \bar{P}(T^n(x)), \ldots)$$

of x to its infinite $T,\bar{P}$-name takes X to Y_P and is Borel in the sense that

$$\mathbf{p}^{-1}(\mathscr{B}) \subseteq \bigvee_{i=-\infty}^{\infty} T^{-i}(\bar{P}) \subseteq \mathscr{F}.$$

As $\mathbf{p}(T(x)) = S(\mathbf{p}(x))$, $\mathbf{p}(\mu) = v_{(\bar{X},\bar{P})}$ is an S-invariant Borel measure on Y_P. Notice that up to completing $\mathscr{B}$ with respect to $v_{(\bar{X},\bar{P})}$, $(Y_P, \mathscr{B}, v_{(\bar{X},\bar{P})}, S)$ is isomorphic to T acting on the factor algebra $\bigvee_{i=-\infty}^{\infty} T^{-i}(\bar{P})$. A pair $(\bar{X}, \bar{P})$ is referred to as a *process* or, more precisely, a *process with state space P*.

Two processes $(\bar{X}_1, \bar{P}_1)$ and $(\bar{X}_2, \bar{P}_2)$ are said to be *identical* if they project to the same measures

$$v_{(\bar{X}_1,\bar{P}_1)} = v_{(\bar{X}_2,\bar{P}_2)}.$$

Thus

$$((Y_P, \mathscr{B}, \nu_{(\bar{X},\bar{P})}, S), P)$$

and

$$(\bar{X}, \bar{P})$$

are identical processes. Said still another way, $(\bar{X}_1, \bar{P}_1)$ and $(\bar{X}_2, \bar{P}_2)$ are identical processes if T_1 restricted to $\bigvee_{i=-\infty}^{\infty} T_1^{-i}(\bar{P}_1)$ and T_2 restricted to $\bigvee_{i=-\infty}^{\infty} T_2^{-i}(\bar{P}_2)$ are isomorphic by a map which simply takes a $T_1,\bar{P}_1$-name to the identical $T_2,\bar{P}_2$-name.

The topology on η_P thus can be regarded as a topology on processes. Two processes are not separated by this topology exactly when they are identical. We write the metric

$$\|\bar{X}_1, \bar{P}_1; \bar{X}_2, \bar{P}_2\| = \|\nu_{(\bar{X}_1,\bar{P}_1)}, \nu_{(\bar{X}_2,\bar{P}_2)}\|, \tag{7.5}$$

and still refer to this as the weak* topology (in probabilistic terms it would be called the weak or vague topology).

We end this section with a small lemma concerning entropy. For notational convenience, we write $(\bar{Y}_P, \nu)$ for the system $(Y_P, \mathscr{B}, \nu, S)$, and $h_\nu(S)$ for its measure-theoretic entropy.

Lemma 7.1 *Suppose $\nu_i \in \eta_P$ and $\nu_i \to \nu$ weak*. Then*

$$\overline{\lim_{i\to\infty}}\, h_{\nu_i}(s) \le h_\nu(s).$$

Thus if $(\bar{X}_i, \bar{P}_i) \to (\bar{X}, \bar{P})$ weak, then*

$$\overline{\lim_{i\to\infty}}\, h(T_i, \bar{P}_i) \le h(T, P).$$

Proof We know, for any $\nu \in \eta_P$, P is a generator for $(\bar{Y}_P, \nu)$. We know

$$h_\nu(S) = \lim_{n\to\infty} h_\nu\left(P \,\middle|\, \bigvee_{j=-1}^{-n} S^{-j}(P)\right)$$

(see Exercise 5.5) and in fact the terms in this limit are non-increasing (Theorem 5.27). For any fixed n,

$$\lim_{n\to\infty} h_{\nu_i}\left(P \,\middle|\, \bigvee_{j=-1}^{-n} S^{-j}(P)\right) = h_\nu\left(P \,\middle|\, \bigvee_{j=-1}^{-n} S^{-j}(P)\right)$$

as these values $h_{\nu_i}(P|\bigvee_{j=-1}^{-n} S^{-j}(P))$ depend only on the ν_i measure of sets in $\bigvee_{j=0}^{-n} S^{-j}(P) \subseteq P_n$.

As $h_{\nu_i}(P|\bigvee_{j=-1}^{-\infty} S^{-j}(P)) \le h_{\nu_i}(P|\bigvee_{j=-1}^{-n} S^{-j}(P))$,

$$\overline{\lim}\, h_{\nu_i}\left(P \,\middle|\, \bigvee_{j=-1}^{-\infty} S^{-j}(P)\right) \le h_{\nu_i}\left(P \,\middle|\, \bigvee_{j=-1}^{-n} S^{-j}(P)\right)$$

for all n. Let $n \to \infty$. ∎

7.2 Painting names on towers and Generic names

We now describe a method for constructing partitions labelled by P. It is the essential tool for creating our representations. We call it *painting a name on a tower*. Suppose we have a finite P-name, i.e., a sequence

$$\mathbf{p} = (p_{i_0}, p_{i_1}, \ldots, p_{i_{N-1}}) \in P^N.$$

Also, in some dynamical system $\bar{X}$ we have a tower of height N consisting of disjoint sets

$$F,\ T(F), \ldots,\ T^{N-1}(F).$$

To paint the name $\mathbf{p}$ on this tower is to define a map $\bar{P}$; $\bigcup_{j=0}^{N-1} T^j(F) \to P$ where $\bar{P}(x) = p_{i_j}$ for $x \in T^j(F)$. Notice $\bar{P}$ does not partition all of X, only the tower.

Suppose we are given a finite collection of names $\mathbf{p}_1, \ldots, \mathbf{p}_s \in \mathbf{P}^N$, and for each a tower with bases $F_1, \ldots, F_s$, and height N, all of which are disjoint. We can paint $\mathbf{p}_j$ onto the tower over F_j giving a partition $\bar{P}$ of the union of all the towers. To actually partition all of X, imagine $\bar{P}$ to have been extended outside the towers in some perhaps arbitrary way. We will describe explicitly how when need be. As we said earlier, this painting of names on towers is the critical tool we will use in the proofs of our representations. There are three parts to the process of painting. We must have names to paint, towers to paint them on and an assignment of names to towers.

We begin with a discussion of names. Suppose $(\bar{X}, \bar{P})$ is a process with state space P. Let $\{B_{i,n}\}_{i=1}^{S(n)}$ be a listing of the elements of $\bigvee_{i=-n}^{n} T^{-i}(\bar{P})$.

Definition 7.1 We say $x \in X$ is ε, *N-generic for* $(\bar{X}, \bar{P})$ if for all n, $0 \le n \le \log_2(2/\varepsilon) + 1$, and all $B_{i,n} \in \bigvee_{i=-n}^{n} T^{-i}(\bar{P})$,

$$\left| \frac{1}{n} \sum_{j=0}^{N-1} \chi_{B_{i,n}}(T^j(x)) - \mu(B_{i,n}) \right| < \frac{1}{2}\left(\frac{\varepsilon}{2} - \frac{2n}{N}\right) \mu(B_{i,n}).$$

Notice that if x is ε, N-generic for $(\bar{X}, \bar{P})$, then any other x' with $\bar{P}(T^j(x)) = \bar{P}(T^j(x'))$, $-\log_2(2/\varepsilon) - 1 \le j \le N + \log_2(2/\varepsilon) + 1$ is also ε, N-generic.

Lemma 7.2 *If $\bar{X}$ is ergodic and $\bar{P}$ is a finite partition, then for any $\varepsilon > 0$, once N_0 is large enough, the set of points which are ε, N-generic for $(\bar{X}, \bar{P})$ for all $N \ge N_0$ has measure at least* $(1 - \varepsilon)$.

Proof Fixing $\varepsilon > 0$, there are only finitely many sets $B_{i,n}$, $0 \le n \le \log_2(2/\varepsilon) + 1$. We need only consider those with $\mu(B_{i,n}) > 0$. Set a lower bound for N_0, $N_0 > 16(\log_2(2/\varepsilon) + 1)/\varepsilon$. Further, require N_0 so large that by the Birkhoff theorem (3.4), for all but a subset of X of measure at most ε,

$$\left|\frac{1}{n}\sum_{j=0}^{N-1} \chi_{B_{i,n}}(T^j(x)) - \mu(B_{i,n})\right| < \frac{\varepsilon}{8}\mu(B_{i,n}) < \frac{1}{2}\left(\frac{\varepsilon}{2} - \frac{2n}{N}\right)\mu(B_{i,n})$$

for all such $B_{i,n}$, for all $N \geq N_0$. ■

Notice that whether or not a point is in a set $B_{i,n}$ is determined by the name

$$P(T^{-n}(x)), \ldots, P(x), \ldots, P(T^n(x)).$$

Thus if we are given some P-name

$$(p_{i_{-n}}, p_{i_{-n+1}}, \ldots, p_{i_0}, \ldots, p_{i_{N-1+n}}),$$

for each index $0 \leq j \leq N$, by reading the name

$$(p_{i_{j-n}}, \ldots, p_{i_{j+n}})$$

we will either see a name corresponding to $B_{i,n}$ or we will not.

Let $\delta(B_{i,n}, \{p_i\}_{j=-n}^{N-1+n})$ be the density in $(0, \ldots, N)$ of occurrences of indices j where $(p_{i_{j-n}}, \ldots, p_{i_{j+n}})$ corresponds to $B_{i,n}$.

Definition 7.2 We say a P,N-name $(p_{i_0}, \ldots, p_{i_{N-1}})$ is *ε-generic for* $(\bar{X}, P)$ if for all $0 \leq n \leq \log_2(2/\varepsilon) + 1$, and any extension of the name

$$p_{i_{-n}}, \ldots, p_{i_0}, \ldots, p_{i_{N-1}}, \ldots, p_{i_{N-1+n}},$$

we have

$$|\delta(B_{i,n}, \{p_i\}_{j=-n}^{N-1+n}) - \mu(B_{i,n})| < \frac{\varepsilon}{4}.$$

Corollary 7.3 *If $x \in X$ is ε,N-generic for $(\bar{X}, P)$, then the T,P,N-name of x, $P(x), \ldots, P(T^{N-1}(x))$ is ε-generic for $(\bar{X}, P)$.* ■

Theorem 7.4 *Suppose we are given a collection of P-names*

$$(P_{i(0,1)}, \ldots, P_{i(N,1)})$$
$$\vdots$$
$$(P_{i(0,t)}, \ldots, P_{i(N,t)})$$

all of which are ε-generic for some $(\bar{X}_1, P_1)$. Suppose we also have a collection of disjoint towers of height N in $\bar{X}_2$ with bases $F_1, \ldots, F_t$ with

$$\mu_2\left(\bigcup_{k=1}^{t}\left(\sum_{j=0}^{N-1} T_2^j(F_j)\right)\right) > 1 - \varepsilon_1.$$

If we paint these names on the corresponding towers we will get a partition P_2 of X_2.

We can conclude

$$\|\bar{X}_1, P_1; \bar{X}_2, P_2\| < \varepsilon + \varepsilon_1.$$

Proof For any $0 \leq n \leq \log_2(2/\varepsilon) + 1$, each set $B^1_{j,n} \in \bigvee_{i=-n}^{n} T_1^{-i}(P_1)$ corresponds to a set

$$B^2_{j,n} \in \bigvee_{i=-n}^{n} T_2^{-i}(P_2)$$

with the same P-name. We can estimate $\mu_2(B^2_{j,n})$ by splitting it into that part within the tower and that outside. For a given tower, say based on F_k, there is a fixed P_2-name up the tower, the name $(p_{i(0,k)}, \dots, p_{i(n,K)})$. This name may extend in various ways as we look n steps before and after the tower. But independent of this, as the name is ε-generic,

$$\left| \frac{\mu_2\left(B^2_{j,n} \cap \bigcup_{i=0}^{N-1} T_2^i(F_k)\right)}{\mu_2\left(\bigcup_{i=0}^{N-1} T_2^i(F_k)\right)} - \mu_1(B^1_{j,n}) \right| < \frac{\varepsilon}{4} \mu_i(B^1_{j,n}).$$

Thus unioning over the towers and summing over the various $B^1_{j,n}$,

$$\sum_j \left| \frac{\mu_1\left(B^1_{j,n} \cap \left(\bigcup_{k=1}^{t} \bigcup_{i=0}^{N-1} T_2^i(F_k)\right)\right)}{\mu_1\left(\bigcup_{k=1}^{t} \bigcup_{i=0}^{N-1} T_2^i(F_k)\right)} - \mu_2(B^2_{j,n}) \right| < \frac{\varepsilon}{4}. \tag{7.6}$$

The points outside the tower can lie in any $B^1_{j,n}$, but only one such. Thus

$$\sum_j |\mu_1(B^1_{j,n}) - \mu_2(B^2_{j,n})| < \frac{\varepsilon}{4} + \varepsilon_1.$$

but as

$$\sum_{k=[\log_2(2/\varepsilon)]+2}^{\infty} 2^{-k} < \frac{\varepsilon}{4},$$

we are done. ■

This theorem explains our rather complex definition of an ε, N-generic point. The important results are Lemma 7.2 and Theorem 7.4. We can jazz up the argument of Theorem 7.4 a little to obtain the fact that ergodic measure are weak* dense in η_p.

Theorem 7.5 *Given any measure $\nu \in \eta_p$ and ergodic, non-periodic system $\bar{X}$, and $\varepsilon > 0$ there is a partition $\bar{P}$ of X with*

$$\|(Y_P, \nu), P; \bar{X}, \bar{P}\| < \varepsilon.$$

Proof First, as η_P is compact, convex and its extreme points are ergodic, we can find a finite convex combination $\nu' = (\sum_{i=1}^{k} \alpha_i \nu_i)$ of ergodic measure ν_i with

$$\|\nu', \nu\| < \frac{\varepsilon}{2}.$$

Using Lemma 7.2, select N_0 large enough and P,N_0-names

$$\mathbf{p}_t = (p_{i(0,t)}, p_{i(1,t)}, \ldots, p_{i(N_0-1,t)}), \quad t = 1, \ldots, k,$$

with $\mathbf{p}_t$ an $\varepsilon/4$-generic name for $((\bar{Y}_P, v_t), P)$.

In $\bar{X}_2$, use the Rohlin lemma (Theorem 3.10) to find a set $F \subseteq X$ with F, $T_2(F)$, ..., $T_2^{N_0-1}(F)$ disjoint and covering all but $\varepsilon/4$ in measure of X_2, Partition F into k pieces, $F_1, \ldots, F_k$ so that

$$\mu_2(F_i)/\mu_2(F) = \alpha_i.$$

Paint $\mathbf{p}_t$ on the tower over R_t. This constructs $\bar{P}$.

Following precisely the computation of Theorem 7.4,

$$\|(\bar{Y}_P, v'), P; \bar{X}, \bar{P}\| \leq \frac{\varepsilon}{4} + \frac{\varepsilon}{4} = \frac{\varepsilon}{2}.$$

We conclude

$$\|(\bar{Y}_P, v), P; \bar{X}, \bar{P}\| < \varepsilon. \qquad \blacksquare$$

Exercise 7.1 We can improve on Theorem 7.5 slightly. Not only are the ergodic measures dense in η_p, they are a G_δ. Consider the subset

$$S(B_{i,n}, \varepsilon, N) = \left\{ v \in \eta_P : \left\| \frac{1}{N} \sum_{j=0}^{N-1} \chi_{B_{i,n}}(S^j(x)) - v(B_{i,n}) \right\|_2 < \varepsilon \right\}.$$

1. Show $S(B_{i,n}, \varepsilon, N)$ is weak* open. Hence so is $S(B_{i,n}, \varepsilon) = \bigcup_{N=1}^{\infty} S(B_{i,n}, \varepsilon, N)$.
2. Show that v is ergodic iff $v \in \bigcap_{i,n,m} S(B_{i,n}, 1/m)$.

Hint: Theorem 3.1 and Corollary 3.16.

Exercise 7.2 Follow the above reasoning to show that the $v \in \eta_P$ for which $(\bar{Y}_P, v)$ is weakly mixing are a dense G_δ. Hint: Remember v will be weakly mixing exactly when $v \times v$ is ergodic for $S \times S$. So show that the measures in $\eta_{P \times P}$ of the form $v \times v$ are a closed convex set and proceed.

Exercise 7.3 Show that the measures v for which $\bar{Y}_P$ is mixing are dense, but meager, i.e., its complement contains a dense G_δ. Hint: Show that the rigid processes (Exercise 4.5) are a dense G_δ.

7.3 $\bar{d}$-Metric and entropy

We now introduce a much stronger metric on processes than the weak*, one intimately connected with joinings. Where the weak* metric says two processes are close if for some long, but finite, period of time they moved sets in approximately the same way, the $\bar{d}$-metric will ask that the two processes look approximately the same forever. Just as with the weak* topology, we can interchangeably view $\bar{d}$ as a metric on processes $(\bar{X}, \bar{P})$ or as a metric on measures in η_P.

For two processes $(\bar{X}_1, \bar{P}_1)$ and $(\bar{X}_2, \bar{P}_2)$, consider the space of joinings $J(\bar{X}_1, \bar{X}_2)$. For any $\hat{\mu} \in J(\bar{X}_1, \bar{X}_2)$ we can compute how closely $\bar{P}_1$ and $\bar{P}_2$ have been matched by $\hat{\mu}(\bar{P}_1 \times X_2 \Delta X_1 \times \bar{P}_2)$, where $\bar{P}_i \times X_j$ indicates the lifting of $\bar{P}_i$ to the joined spaces.

Definition 7.3

$$\bar{d}(\bar{X}_1, \bar{P}_1; \bar{X}_2, \bar{P}_2) = \inf_{\hat{\mu} \in J(\bar{X}_1, \bar{X}_2)} \hat{\mu}(\bar{P}_1 \times X_2 \Delta X_1 \times \bar{P}_2).$$

We will write

$$\bar{d}(\nu_1, \nu_2) \quad \text{for} \quad \bar{d}((\bar{Y}_P, \nu_1), P; (\bar{Y}_P, \nu_2), P).$$

Lemma 7.6

1. *$\bar{d}$ is a metric on equivalence classes of identical processes.*
2. *The infimum of the definition of $\bar{d}$ is actually a minimum.*
3. *If $\bar{X}_1$ and $\bar{X}_2$ are ergodic, then the $\bar{d}$-distance between $(\bar{X}_1, \bar{P}_1)$ and $(\bar{X}_2, \bar{P}_2)$ is achieved by an ergodic joining.*

Proof 1. Certainly $\bar{d}(\bar{X}_1, \bar{P}_1; \bar{X}_2, \bar{P}_2) = 0$ iff the two processes are identical. Symmetry is obvious. The triangle inequality follows via the construction of the relatively independent joining. If $\hat{\mu}_1 \in J(\bar{X}_1, \bar{X}_2)$ and $\hat{\mu}_2 \in J(\bar{X}_2, \bar{X}_3)$, then we can construct $\hat{\mu}$, a joining of $\bar{X}_1, \bar{X}_2$ and $\bar{X}_3$ simultaneously as the relatively independent joining of $\hat{\mu}_1$ and $\hat{\mu}_2$ over the common factor $\bar{X}_2$. Certainly

$$\begin{aligned}
&\hat{\mu}(\bar{P}_1 \times \bar{X}_2 \times \bar{X}_3 \Delta X_1 \times X_2 \times \bar{P}_3) \\
&\quad \leq \hat{\mu}(\bar{P}_1 \times X_2 \times X_3 \Delta X_1 \times \bar{P}_2 \times X_3) \\
&\qquad + \hat{\mu}(X_1 \times \bar{P}_2 \times X_3 \Delta X_1 \times X_2 \times \bar{P}_3) \\
&\quad = \hat{\mu}(\bar{P}_1 \times X_2 \Delta X_1 \times \bar{P}_2) + \hat{\mu}_2(\bar{P}_2 \times X_3 \Delta X_2 \times \bar{P}_3),
\end{aligned}$$

and as $\hat{\mu}$, restricted to the first and third coordinates is in $J(\bar{X}_1, \bar{X}_3)$ we are done.

2 and 3. Remember $J(\bar{X}_1, \bar{X}_2)$ is weak*-compact and convex. The function $V(\hat{\mu}) = \hat{\mu}(\bar{P}_1 \times X_2 \Delta X_1 \times \bar{P}_2)$ is weak*-continuous and linear. Hence it has a minimum achieved on the boundary of $J(\bar{X}_1, \bar{X}_2)$. By Theorem 6.3, if $\bar{X}_1$ and $\bar{X}_2$ are ergodic this boundary consists of the ergodic joinings. ∎

Theorem 7.7 *For any $\varepsilon > 0$, and processes $(\bar{X}_1, \bar{P}_1), (\bar{X}_2, \bar{P}_2)$, if*

$$\bar{d}(\bar{X}_1, \bar{P}_1; \bar{X}_2, \bar{P}_2) < \frac{\varepsilon}{4 \log_2(2/\varepsilon) + 4}$$

then $\|\bar{X}_1, \bar{P}_1; \bar{X}_2, \bar{P}_2\| < \varepsilon$.

Proof Let $\hat{\mu}$ achieve the $\bar{d}$-distance. For $n \le \log_2(2/\varepsilon) + 1$,

$$\hat{\mu}\left(\bigvee_{i=-n}^{n} T_1^{-i}(\bar{P}_1) \times X_2 \,\Delta\, X_1 \times \bigvee_{i=-n}^{n} T_2^{-i}(\bar{P}_2)\right)$$
$$= \hat{\mu}\left(\bigvee_{i=-n}^{n} (T_1^{-i}(\bar{P}_1) \times X_2 \,\Delta\, X_1 \times T_2^{-i}(\bar{P}_2))\right)$$
$$\le (2n+1)\mu(\bar{P}_1 \times X_2 \,\Delta\, X_1 \times \bar{P}_2) < \frac{\varepsilon}{2}.$$

Thus for such n

$$\frac{1}{2}\sum_i |\mu_1(B^1_{i,n}) - \mu_2(B^2_{i,n})| < \frac{\varepsilon}{2}.$$

As

$$\sum_{i > \log_2(2/\varepsilon)+1} 2^{-i} < \frac{\varepsilon}{2},$$

we are done. ■

Thus the $\bar{d}$-metric is a strengthening of the weak* metric. It is in fact an extreme strengthening. We can see this in a variety of ways. First, η_P is *not* $\bar{d}$-compact. Consider the uncountable collection of pairwise disjoint maps of Exercise 6.2. Each is equipped with a generating partition $P_{\mathbf{e}}$ into two sets (the first tower and the spacers). These sets have measures 2/3 and 1/3, respectively. As the only joining of two of these is product measure, $\bar{d}(\bar{X}_{\mathbf{e}_1}, P_{\mathbf{e}_1}; \bar{X}_{\mathbf{e}_2}, P_{\mathbf{e}_2}) = 5/9$. Thus in η_P there is a collection of cardinality the continuum whose elements are pairwise 5/9 apart in $\bar{d}$. Similar reasoning leads to the conclusion that no K-system can be the $\bar{d}$-limit of zero entropy systems.

A third indication, intimately related to the above remark, is the following two theorems.

Theorem 7.8 *The ergodic measures in η_P are $\bar{d}$-closed.*

Proof Suppose $\nu_i \to \nu$ in $\bar{d}$. Choose i so large that

$$\bar{d}(\nu_i, \nu) < \varepsilon^2,$$

and is achieved by a joining, $\hat{\mu}$. Let $\hat{\mu} = \int_0^1 \hat{\mu}_t \, dt$ be its ergodic decomposition. By the same reasoning as Theorem 6.3, a.e. $\hat{\mu}_t$ has first marginal ν_i. Letting ν_t be its second marginal,

$$\nu = \int_0^1 \nu_t \, dt$$

is an ergodic decomposition of ν.

As

$$\bar{d}(v_i, v) = \int_0^1 \bar{d}(v_i, v_t)\,\mathrm{d}t < \varepsilon^2,$$

for all but ε in measure of the v_t, $d(v, v_t) < \varepsilon$. This says for a.e. t, $\bar{d}(v, v_t) = 0$ or $v = v_t$, hence v is ergodic. ■

Theorem 7.9 *If* $(\bar{X}_1, \bar{P}_1)$ *and* $(\bar{X}_2, \bar{P}_2)$ *are ergodic and*

$$\bar{d}(\bar{X}_1, \bar{P}_1; \bar{X}_2, \bar{P}_2) < \varepsilon,$$

then

$$h(T_1, \bar{P}_1) \geq h(T_2, \bar{P}_2) - H(\varepsilon) - \varepsilon \log_2(n). \tag{7.7}$$

Proof Let $\hat{\mu}$ be an ergodic joining that achieves the $\bar{d}$-distance. As $\hat{\mu}(\bar{P}_1 \times X_2 \Delta X_1 \times \bar{P}_2) < \varepsilon$, the result follows from Lemma 5.10. ■

Thus entropy is in fact $\bar{d}$-continuous. The $\bar{d}$-distance is intimately related to what is referred to in information theory as the Hamming distance between names.

Definition 7.4 If $\mathbf{p}_1$ and $\mathbf{p}_2$ are two names in P^N, then the *Hamming distance* between them is

$$\|\mathbf{p}_1, \mathbf{p}_2\|_H^N = \#\{i : p_1(i) \neq p_2(i)\}.$$

the $\bar{d}_N$-distance is the normalized Hamming distance, i.e.,

$$\bar{d}_N(\mathbf{p}_1, \mathbf{p}_2) = \frac{\#\{i : p_1(i) \neq p_2(i)\}}{N}.$$

Theorem 7.10 *Suppose* $(\bar{X}_1, \bar{P}_1)$ *and* $(\bar{X}_2, \bar{P}_2)$ *are ergodic processes,* $N_k \nearrow \infty$ $\varepsilon_k \searrow 0$ *and we have sequences of names*

$$\mathbf{p}_k^1 \in P^{N_k}$$

and

$$\mathbf{p}_k^2 \in P^{N_k},$$

and $\mathbf{p}_k^j$ *is* ε_k*-generic for* $(\bar{X}_j, \bar{P}_j)$, $j = 1, 2$. *Then*

$$\bar{d}(\bar{X}_1, \bar{P}_1; \bar{X}_2, \bar{P}_2) \leq \varliminf_{k\to\infty} \bar{d}_{N_k}(\mathbf{p}_k^1, \mathbf{p}_k^2).$$

Proof For each k, consider the double name

$$\mathbf{p}_k = ((p_{(0,k)}^1, p_{(0,k)}^2), \ldots, (p_{(N_k-1,k)}^1, p_{(N_k-1,k)}^2)) \in (P \times P)^{N_k}.$$

Let $(\bar{X}_3)$ be some arbitrary non-periodic ergodic system. Build a tower in $\bar{X}_3$ with base F_k, and height N_k covering all but ε_k of X_3.

Paint $\mathbf{p}_k$ on this tower, constructing a partition $\tilde{P}_1^k \vee \tilde{P}_2^k$. By Theorem 7.4

$$\|(\bar{X}_1, \bar{P}_1), (\bar{X}_3, \tilde{P}_1^k)\| \leq 2\varepsilon_k$$

and

$$\|(\bar{X}_2, \bar{P}_2), (\bar{X}_3, \tilde{P}_2^k)\| \leq 2\varepsilon_k.$$

Furthermore,

$$\mu_3(\tilde{P}_1^k \,\Delta\, \tilde{P}_2^k) \leq \frac{1}{N_k} \|\mathbf{p}_k^1, \mathbf{p}_k^2\|_H^{N_k} + \varepsilon_k.$$

Consider the sequence of measures

$$\nu_{(\bar{X}_3, \tilde{P}_1^k \vee \tilde{P}_2^k)} = \nu_k \in \eta_{P_\alpha \times P_\alpha}.$$

Let ν be the limit of a weak* convergent subsequence on which

$$\underline{\lim}\, \frac{1}{N_k} \|\bar{p}_k^1, \bar{p}_k^2\|_H^{N_k}$$

is a limit. Thus $(\bar{X}_1, \bar{P}_1)$ and $(Y_{P_1 \times P_2}, \nu, P_1)$ are identical, as are $(\bar{X}_2, \bar{P}_2)$ and $(Y_{P_1 \times P_2}, \nu, P_2)$. Hence ν entends to a joining of $\bar{X}_1$ and $\bar{X}_2$, and

$$\bar{d}(\bar{X}_1, \bar{P}_1; \bar{X}_2, \bar{P}_2) \leq \nu(P_1 \,\Delta\, P_2) = \underline{\lim}\, \bar{d}_{N_k}(\mathbf{p}_k^1, \mathbf{p}_k^2). \qquad ■$$

Corollary 7.11 *Suppose* $(\bar{X}_1, \bar{P}_1)$ *and* $(\bar{X}_2, \bar{P}_2)$ *are ergodic processes, and* $N_k \nearrow \infty$. *Suppose we have a sequence of subsets* $A_k \in X_1$, $\mu(A_k) > \alpha > 0$, *and measure-preserving maps* $\varphi_k : A_k \to X_2$. *Then*

$$\bar{d}(\bar{X}_1, \bar{P}_1; \bar{X}_2, \bar{P}_2) \leq \underline{\lim}\, \frac{1}{\mu_1(A_k)} \int_{A_k} \bar{d}_{N_k}(\mathbf{p}_{N_k}^1(x), \mathbf{p}_{N_k}^2(\varphi_k(x)))\, d\mu_1.$$

Proof Since $N_k \nearrow \infty$, we can find $\varepsilon_k \searrow 0$ by Lemma 7.2 so that all but ε_k of the $x \in X_j$ are ε_k, N_k-generic for X_j, $j = 1$ or 2. Once $2\varepsilon_k < \alpha$, there must be points $x_k \in S_k$ with both $\mathbf{p}_{N_k}^1(x)$, ε_k, N_k-generic for $(\bar{X}_1, \bar{P}_1)$ and $\mathbf{p}_{N_k}^2(\varphi_k(x))$, ε_k, N_k-generic for $(\bar{X}_2, \bar{P}_2)$ and

$$\bar{d}_{N_k}(\mathbf{p}_{N_k}^1(x_k), \mathbf{p}_{N_k}^2(\varphi_k(x_k))) \leq \left(\frac{1}{\mu_1(A_k)} \int_{A_k} \bar{d}_{N_k}(\mathbf{p}_{N_k}^1(x), \mathbf{p}_{N_k}^2(\varphi_k(x)))\, d\mu_1\right)\left(1 - \frac{2\varepsilon_k}{\alpha}\right).$$

Apply Theorem 7.10 to finish the result. ■

When we apply Corollary 7.11 later, the sets A_k will be all of X_1. What is important here is that the maps φ_k pairing points in X_1 to points in X_2 need not be joinings, i.e., need not commute with the transformations.

Exercise 7.4 Show that the weakly mixing processes are $\bar{d}$-closed.

Exercise 7.5 Show that the mixing and K-processes are $\bar{d}$-closed.

Exercise 7.6 Show that if $\bar{X}_1$ and $\bar{X}_2$ have $\bar{d}(\bar{X}_1, \bar{P}_1; \bar{X}_2, \bar{P}_2) = \bar{d}$, then for a.e. $x \in X_1$, there is a sequence of names $\mathbf{p}_n^2(x) = \{p^2_{i(0,n)}(x), p^2_{i(1,n)}(x), \ldots, p^2_{i(n,n)}(x)\} \in P^n$ which becomes ever more generic for $(\bar{X}_2, \bar{P}_2)$ and for which

$$\lim_{n\to\infty} \frac{1}{n} \| \mathbf{p}_n(x), \mathbf{p}_n^2 \|_H^n = \bar{d}.$$

We end this section with a very important perturbation argument. We saw in Theorem 7.9 that entropy is $\bar{d}$-continuous. Hence a small $\bar{d}$-perturbation cannot change entropy by much. What we want to see now is that for $v \in \eta_P$, as long as $h_v(x) < \log_2(n)$, i.e., v is not the Bernoulli n-shift, we can perturb some entropy into v. We need a particularly strong version of this, we define a strengthening of the $\bar{d}$ metric.

Definition 7.5 Set

$$\bar{d}^N(\bar{X}_1, \bar{P}_1; \bar{X}_2, \bar{P}_2) = \bar{d}\left(\bar{X}_1, \bigvee_{i=-N}^{N} T_1^{-i}(\bar{P}_1); \bar{X}_2, \bigvee_{i=-N}^{N} T_2^{-i}(\bar{P}_2) \right),$$

i.e., we measure not just how closely $\bar{P}_1$ can be joined to $\bar{P}_2$, but how closely $\bigvee_{i=-N}^{N} T_1^{-i}(\bar{P}_1)$ can be joined to $\bigvee_{i=-N}^{N} T_2^{-i}(\bar{P}_2)$.

It is a computation that

$$\bar{d}^N(\bar{X}_1, \bar{P}_1; \bar{X}_2, \bar{P}_2) \le N\bar{d}(X_1, \bar{P}_1; X_2, \bar{P}_2).$$

Exercise 7.7 Show that for any $\varepsilon > 0$, for $N > \log_2(2/\varepsilon) + 1$

$$\|v_1, v_2\| \le \bar{d}^N(v_1, v_2) + \varepsilon.$$

Theorem 7.12 *For any $v_1 \in \eta_P$, $\varepsilon > 0$ and natural number N, there is an $v_2 \in \eta_P$ so that*

(1) $h_{v_2}(S) \ge h_{v_1}(S) + \varepsilon(\log_2(n) - h_{v_1}(S))$; *and*

(2) $\bar{d}^N(v_1, v_2) < 2\varepsilon$. (7.8)

Further, if v_1 is ergodic, so is v_2.

Proof Let v_0 be the Bernoulli n-shift $(n^{-1}, n^{-1}, \ldots, n^{-1})$ with $h_{v_0}(S) = \log_2(n)$, the unique maximal entropy measure in η_P. Let $(\bar{X}, R)$ be any weakly mixing process, R, a partition into two sets r_1, r_2 where

$$\mu\left(\bigcap_{i=-N}^{N} T^{-i}(r_1) \right) \ge (1 - \varepsilon)\mu(r_1)$$

and $\mu(r_2) = \varepsilon$. Such a partition can be found in any non-periodic ergodic $\bar{X}$ using the Rohlin lemma.

Consider the space $Z = Y_p \times X \times Y_p$ with measure $v_1 \times \mu \times v_0$, and transformation $S \times T \times S$. Construct a partition of $\hat{Y}$ as follows

$$\bar{P}(y_1, x, y_2) = \begin{cases} P(y_1) & \text{if } x \in r_1 \\ P(y_2) & \text{if } x \in r_2. \end{cases}$$

Let $\nu_2 \in \nu_{(\bar{Z},\bar{P})} \in \eta_P$.

To see (2) just notice that ν_1 and ν_2 are joined in $\bar{Z}$ as $P(y_1)$ and $\bar{P}(y_1, x, y_2)$. In Z, if $x \in \bigcap_{i=-N}^{N} T^{-i}(r_1)$, then (y_1, x, y_2) and y_1 belong to the same elements of $\bigvee_{i=-N}^{N}(S \times T \times S)^{-i}(\bar{P})$ and $\bigvee_{i=-N}^{N} S^{-i}(P)$, respectively. Thus

$$\bar{d}^N(\nu_1, \nu_2) \le 1 - \mu\left(\bigcap_{i=-N}^{N} T^{-i}(r_1)\right) \le 2\varepsilon.$$

For (1) we remember

$$\begin{aligned}
h_{\nu_2}(S, P) &= h_{\nu_1 \times \mu \times \nu_0}(S \times T \times S, \bar{P}) \\
&= \int I\left(\bar{P} \middle| \bigvee_{i=-1}^{-\infty} (S \times T \times S)^{-i}(\bar{P})\right) d(\nu_1 \times \mu \times \nu_0) \\
&\ge \int I\left(\bar{P} \middle| \bigvee_{i=-1}^{-\infty} S^{-i}(P) \times \bigvee_{i=0}^{-\infty} T^{-i}(R) \times \bigvee_{i=-1}^{-\infty} S^{-i}(P)\right) d\nu_1 \times \mu \times \nu_0 \\
&= \int_{Y \times r_2 \times Y} I\left(Y \times X \times P \middle| \bigvee_{i=-1}^{-\infty} S^{-i}(P)\right. \\
&\qquad \left. \times \bigvee_{i=0}^{-\infty} T^{-i}(R) \times \bigvee_{i=-1}^{-\infty} S^{-i}(P)\right) d(\nu_1 \times \mu \times \nu_0) \\
&\quad + \int_{Y \times r_1 \times Y} I\left(P \times X \times Y \middle| \bigvee_{i=-1}^{-\infty} S^{-i}(P)\right. \\
&\qquad \left. \times \bigvee_{i=0}^{-\infty} T^{-i}(R) \times \bigvee_{i=-1}^{-\infty} S^{-i}(P)\right) d(\nu_1 \times \mu \times \nu_0) \\
&= \int_{r_2 \times Y} I\left(P \middle| \bigvee_{i=-1}^{-\infty} S^{-i}(P)\right) d(\mu \times \nu_0) \\
&\quad + \int_{Y \times r_1} I\left(P \middle| \bigvee_{i=-1}^{-\infty} S^{-i}(P)\right) d(\nu_1 \times \mu) \\
&= \mu(r_2)\log_2(n) + \mu(r_1) h_{\nu_1}(S) \\
&= \varepsilon \log_2(n) + (1 - \varepsilon) h_{\nu_1}(S).
\end{aligned}$$

■

7.4 Pure columns and Ornstein's fundamental lemma

Ornstein's fundamental lemma is the critical piece of information we need for our proofs of the Ornstein and Krieger theorems. It is a painting argument. To this point our painting arguments have been rather robust, just painting a name or collection of names on essentially any tower. Here though we will be given some pre-existing partitions and will wish to repaint them to improve their character.

Our first step is to understand how a pre-existing partition paints a tower. Let $(\bar{X}, \bar{P})$ be an ergodic process, and F, $T(F)$, ..., $T^{N-1}(F)$ be a tower in X. The T,P,N-names of points in F cut this tower into subtowers. Suppose $\mathbf{p} = (p_{i_0}, p_{i_1}, \ldots, p_{i_{N-1}}) \in P^N$ is some particular P,N-name. Let $F_{\mathbf{p}} \subset F$ consist of those points with $\mathbf{p}_N(x) = \mathbf{p}$. Such sets partition F, and are bases for disjoint towers. Such a tower over $F_{\mathbf{p}}$ is called a *pure P-column* because each little piece $T^j(F_{\mathbf{p}}) \cap T^j(F)$ lies in a single, pure set of $\bar{P}$, $0 \leq j < N$. Notice that $\bar{P}$ is simply a painting of the names $\mathbf{p}$ on the columns $F_{\mathbf{p}}$.

We want to prove tower-related versions of the ergodic theorem, and the Shannon–McMillan–Breiman theorem. Both these results will rest on the same trick. Even though the base set F occupies only a fraction $1/N$ of the tower, thickening it to $\bigcup_{i=0}^{[\alpha N]} T^i(F)$ occupies a fraction α. Further, the name of a point in the thickening differs in a controlled way from the name of the point below it in F.

Theorem 7.13 *Suppose $(\bar{X}, \bar{P})$ is an ergodic process. For any $\varepsilon > 0$, once N is large enough in any tower F, $T(F), \ldots, T^{N-1}(F)$ with*

(1) $\mu\left(\bigcup_{i=0}^{N-1} T^i(F)\right) > \varepsilon$; *we can conclude*

(2) $\mu(\{x \in F_{\mathbf{p}} : \mathbf{p}$ *is ε-generic for* $(\bar{X}, \bar{P})\}) > (1 - \varepsilon)\mu(F)$.

Proof Notice that if x is $\bar{\varepsilon}, M$-generic for $(\bar{X}, \bar{P})$, then $T^{-i}(x)$ is $(\bar{\varepsilon} + i/(m + 1))$, $(M + i)$-generic for $(\bar{X}, \bar{P})$. Let $\bar{\varepsilon} = \varepsilon^3/8$. Choose M_0 by Lemma 7.2 so that for all but $\bar{\varepsilon}$ of the $x \in X$, x is $\bar{\varepsilon}, M$-generic for *all* $M \geq M_0$.

Let $N(1 - \varepsilon/2) > M_0$ and consider the set of points $B = \bigcup_{i=0}^{[\varepsilon N/2]} T^i(F)$. As $\mu(B) > \varepsilon^2/2$ there is a subset $B_0 \subseteq B$, $\mu(B_0) > (1 - \varepsilon/4)\mu(B)$, and all $x \in B_0$ are $\bar{\varepsilon}, M$-generic for all $M \geq M_0$.

Thus if $x \in B_0 \cap T^i(F)$, $0 \leq i \leq [\varepsilon N/2]$, then x is $\bar{\varepsilon}(N - i)$-generic. Hence $T^{-i}(x) \in F$ is $(\bar{\varepsilon} + i/(m + i))$, N-generic, hence ε, N-generic for $(\bar{X}, \bar{P})$. As $\mu(\bigcup_{i=0}^{[\varepsilon N/2]} T^{-i}(B_0) \cap F) > (1 - \varepsilon/4)\mu(F)$ we are done. ∎

Lemma 7.14 *Suppose $(\bar{X}, \bar{P})$ is an ergodic process. For any $\varepsilon > 0$, if N is large enough, and A is any set with $\mu(A) \geq \varepsilon$, there is, then, a subset $A_0 \subseteq A$, $\mu(A_0) > (1 - \varepsilon)\mu(A)$ and for any $x \in A_0$,*

$$\left| -\frac{1}{N} \log_2(\mu(\mathbf{p}_N(x) \cap A)/\mu(A)) - h(T, P) \right| < \varepsilon.$$

Proof We establish upper and lower estimates separately. First, it is enough to show

$$\left|-\frac{1}{N}\log_2(\mu(\mathbf{p}_N(x)\cap A)) - h(T,P)\right| < \frac{\varepsilon}{2} \tag{7.9}$$

as

$$-\frac{1}{N}\log_2(\mu(A)) < -\frac{\log_2(\varepsilon)}{N} < \frac{\varepsilon}{2}$$

once $N > -2\log_2(\varepsilon)/\varepsilon$.

By the Shannon–McMillan–Breiman Theorem (5.3), once N is large enough, for all but at most $\varepsilon^2/4$ of X,

$$\left|-\frac{\log_2(\mu(\mathbf{p}_N(x)))}{N} - h(T,\bar{P})\right| < \frac{\varepsilon}{2}. \tag{7.10}$$

So certainly $-\log_2(\mu(\mathbf{p}_N(x)) \cap A)/N < h(T,P) + \varepsilon/2$ for all but a fraction $\varepsilon/4$ of A.

For the other inequality, we remind ourselves that by Corollary 5.2 once N is large enough a collection of at most $2^{(h(T,P)+\varepsilon/4)N}$ $T,\bar{P},N$-names covers all but $\varepsilon^2/8$ of X. Such a collection will cover all but a fraction $\varepsilon/8$ of A.

Let $B \subseteq A$ be those $x \in A$ with $\mu(\mathbf{p}_N(x) \cap A) \leq 2^{-(h(T,P)-\varepsilon/2)N}$, i.e., where the lower estimate fails. We conclude

$$\mu(B) \leq \frac{\varepsilon}{8}\mu(A) + 2^{(h(T,P)+\varepsilon/4)N}2^{-(h(T,P)+\varepsilon/2)N} = \frac{\varepsilon}{8}\mu(A) + 2^{-(\varepsilon/4)N}.$$

Once $N > -4\log_2(\varepsilon/8)/\varepsilon$, $\mu(B) < \mu(A)/4$. This gives us both estimates outside a subset of A of measure $\leq \varepsilon\mu(A)/2$. ■

Theorem 7.15 *Suppose $(\bar{X},\bar{P})$ is an ergodic process and $\varepsilon > 0$. Once N is large enough, for any tower $F, T(F), \ldots, T^{N-1}(F)$ with*

(1) $\mu(\bigcup_{i=0}^{N-1} T^i(F)) > \varepsilon$,

we can find a set I of P,N-names so that

(2) $\mu(\bigcup_{\mathbf{p}\in I} F_{\mathbf{p}}) > (1-\varepsilon)\mu(F)$; *and*

(3) *for* $\mathbf{p} \in I$, $|-1/N\log_2(\mu(F_{\mathbf{p}})/\mu(F)) - h(T,\bar{P})| < \varepsilon$.

Proof As in Lemma 7.14, we establish upper and lower estimates separately. Since

$$-\frac{\log_2\mu(F)}{N} \leq -\frac{\log(\varepsilon)}{N} + \frac{\log(N)}{N} < \frac{\varepsilon}{2}$$

once N is large enough, it is enough to show

$$\left|-\frac{1}{N}\log_2(\mu(F_{\mathbf{p}})) - h(T,P)\right| < \frac{\varepsilon}{2}. \tag{7.11}$$

Consider the two sets $A^{(+)} = \bigcup_{i=0}^{[\varepsilon/8N]} T^i(F)$ and $A^{(-)} = \bigcup_{i=-[\varepsilon/8N]}^{0} T^i(F)$. Both have measure at least $\varepsilon^2/8$, independent of N.

For the upper estimate, by Lemma 7.14, if N is large enough, a subset $A_0^{(+)}$ of all but a fraction $\varepsilon/4$ of $A^{(+)}$ is covered by $T,\mathbf{P},[N(1+\varepsilon/4))$-names whose intersections with $A^{(+)}$ have measure at most

$$2^{-(h(T,P)-\varepsilon/4)([N(1-\varepsilon/4)]+1)}.$$

For any $x \in F$ and $0 \le i \le [\varepsilon N/8]$,

$$T^i(\mathbf{p}_N(x) \cap F) \subseteq (\mathbf{p}_{[N(1-\varepsilon/8)]+1}(T^i(x)) \cap A^{(+)}).$$

Thus if $T^i(x) \in A_0^{(+)}$ then $\mu(F_{\mathbf{p}_N(x)}) = \mu(\mathbf{p}_N(x) \cap F) \le 2^{-(h(T,P)-\varepsilon/2)N}$. Such x must cover all but a fraction $\varepsilon/4$ of F.

For the lower estimate, again using Lemma 7.14, if N is large enough there is a subset $A_0^{(-)}$ of all but a fraction $\varepsilon/16$ of $A^{(-)}$, covered by at most

$$2^{-(h(T,P)+\varepsilon/16)([N(1+\varepsilon/8)]+1)}$$

$T,P,[N(1+\varepsilon/8)]+1$-names.

Let $F_0 \subseteq F$ be those points with $T^{-i}(x) \in A_0^{(-)}$ for some $0 \le i \le [\varepsilon N/8]$. Thus $\mu(F_0) \ge (1-\varepsilon/16)\mu(F)$. Each $T,P,[N(1+\varepsilon/8)]+1$-name in $A_0^{(-)}$ gives rise to a possible $[\varepsilon N/8]+1$ different T,P,N-names in F_0, one for each $0 \le i \le [\varepsilon N/8]$. Thus F_0 is covered by at most $([\varepsilon N/8]+1)2^{(h(T,P)+3\varepsilon/16)(n+1)}$ T,P,N-names. Be sure N is large enough that this is less than $2^{(h(T,P)+\varepsilon/4)N}$.

What is the measure of the set of $x \in F_0$ with $\mu(\mathbf{p}_N(x) \cap F) \le 2^{-(h(T,P)+\varepsilon/2)N}$? At most

$$2^{(h(T,P)+\varepsilon/4)-(h(T,P)+\varepsilon/2)N} \le 2^{-\varepsilon/2N}.$$

Being sure $N > 2/\varepsilon \log_2(4N/2)$, and we conclude

$$2^{-\varepsilon/2N} \le \frac{\varepsilon}{4}\mu(F).$$

Thus for all but a fraction $\varepsilon/2$ of F we get the lower estimate and we are done. ■

The next exercise will in fact be used later in the proof of the isomorphism theorem.

Exercise 7.8 In Theorem 7.12 we saw entropy could be perturbed into a process. Here we show it can be perturbed out. Explicitly, suppose $(\bar{X}, \bar{Q})$ is an ergodic non-periodic process and $1 > \alpha > \varepsilon > 0$. Show there is a partition $\bar{Q}_0$ with

(1) $\mu(\bar{Q}_0 \,\Delta\, \bar{Q}_0) < \alpha$; and

(2) $h(T, \bar{Q}_0) < (1-\alpha)h(T, \bar{Q}) + \varepsilon$.

Hint: Construct a sequence of partitions $\bar{Q}_N$ by building a Rohlin tower of height N, covering all but (ε/N) of X. Split the base $F = F_1 \cup F_2$ where $\mu(F_1) = \alpha$. Repaint the tower over F_1 to be all in q_1. Use Theorems 7.4 and 7.10 to show $v_{(\bar{X},\bar{Q}_N)} = v_N \in \eta_Q$ converges weak* to $\alpha v_0 + (1-\alpha)v_{(\bar{X},\bar{Q})}$ where v_0 is a point mass on the sequence of all q_1's. Use Lemma 7.1 to complete things.

We are now prepared to prove Ornstein's fundamental lemma. We will in fact prove two forms, dividing the argument first into a painting project and second a perturbing to gain entropy. It is historically incorrect to call either of these Ornstein's fundamental lemma (see Shields 1973). Embodied in them though is Ornstein's original 'copying' or 'painting' argument, which remains the core insight to the proofs of the Krieger and Ornstein results. The structure of these theorems though has been greatly refined and clarified, most particularly by Kieffer, and Burton and Rothstein. We are presenting this refined version.

Suppose $(\bar{X}_1, \bar{P})$ and $(\bar{X}_2, \bar{Q})$ are two ergodic processes, and $\hat{\mu}$ is an ergodic joining. The question the fundamental lemma addresses is when is it possible to perturb $\hat{\mu}$ only slightly weak* so that $\bar{P}$ becomes approximately $\bar{X}_2$ measurable and $\bar{Q}$ becomes approximately $\bar{X}_1$ measurable. There will obviously be entropy constraints. Furthermore, both processes cannot remain fixed. One of them, $(\bar{X}_1, \bar{P})$, must also be perturbed.

Definition 7.6 We say a partition P is *μ,ε-contained* in a σ-algebra $\mathscr{A}$, written $P \overset{\varepsilon}{\underset{\mu}{\subset}} \mathscr{A}$, if there is a partition $P_1 \subset \mathscr{A}$ with $\mu(P \,\Delta\, P_1) < \varepsilon$.

Lemma 7.16 *Suppose $(X, \bar{P})$ is a ergodic, non-atomic process. For any $\varepsilon > 0$ there is an N_0 and partition P' of x with*

(1) $\mu(\bar{P} \,\Delta\, \bar{P}_1) < \varepsilon$; *and*

(2) $\mu(\{x : \bar{P}'(T^i(x)) = p_1$ *for* $i = 0, 1, \ldots, N_0 - 1\}) = 0$.

Proof As $(X, \bar{P})$ is non-atomic, $\lim_{N\to\infty} \mu(\{x : \bar{P}(T^i(x)) = p_1$ for $i = 0, 1, \ldots N-1\}) = 0$. Choose N_0 so that this measure is less than ε.

Let

$$\bar{P}'(x) = \begin{cases} \bar{P}(x) & \text{if } P_i(x), \quad i = 0, 1, \ldots, N_0 - 1 \text{ are not all } p_i; \\ p_2 & \text{if they are.} \end{cases}$$ ∎

Theorem 7.17 (Ornstein's fundamental lemma, first form) *Suppose $(\bar{X}_1, \bar{P})$ and $(\bar{X}_2, \bar{Q})$ are non-periodic ergodic processes, and*

$$h(T_1, \bar{P}) > h(T_2, \bar{Q}).$$

Further, suppose $\hat{\mu} \in J(\bar{X}_1, \bar{X}_2)$ *is ergodic and* $\varepsilon > 0$. *There is, then, a partition* $\tilde{P}$ *of* X_2 *so that*

(1) $\|(\bar{X}_1 \times \bar{X}_2, \hat{\mu}), \bar{P} \times \bar{Q}; \bar{X}_2, \tilde{P} \vee \bar{Q}\| < \varepsilon$; *and*

(2) $\bar{Q} \overset{\varepsilon}{\underset{\mu_2}{\subset}} \bigvee_{i=-\infty}^{\infty} T_2^{-i}(\tilde{P})$. (7.12)

Proof Using Lemma 7.16, Lemma 5.10 and Theorem 7.7, we can assume without loss of generality that for some N_0, $\mu_1(\{x : \bar{P}(T^i(x)) = p_1 \text{ for } i = 1, 2, \ldots, N_0 - 1\}) = 0$. We begin by applying Theorems 7.13 and 7.15 to our joined process $((\bar{X}_1 \times \bar{X}_2, \hat{\mu})\bar{P} \times \bar{Q})$ and its two coordinate processes $(\bar{X}_1, \bar{P})$ and $(\bar{X}_2, \bar{Q})$.

Let $h(T_1, \bar{P}) - h(T_2, \bar{Q}) > \alpha > 0$ where $\alpha \leq 1$. Be sure N is so large that for any tower in the joined process of height N, covering at least $\varepsilon\alpha/16$ of $X_1 \times X_2$ we have

(1) All but a fraction $\varepsilon\alpha/16$ of the $\bar{P} \times \bar{Q}, N$-names in the base of the tower are $\varepsilon\alpha/16$-generic for $\hat{\mu}$. (7.13)

Note: if a $\bar{P} \times \bar{Q}, N$-name is $\varepsilon\alpha/16$ generic for $\hat{\mu}$, then the coordinate $\bar{P}, N$-names and $\bar{Q}, N$-names are $\varepsilon\alpha/16$-generic for μ_1 and μ_2 respectively.

(2) All but a fraction $\varepsilon\alpha/16$ of the points $(x_1, x_2) \in F$ satisfy

(a) $\left| -\frac{1}{N} \log_2(\hat{\mu}(F_{\mathbf{p}_N(x_1)})/\hat{\mu}(F)) - h(T_1, \bar{P}) \right| < \frac{\varepsilon\alpha}{16}$,

(b) $\left| -\frac{1}{N} \log_2(\hat{\mu}(F_{\mathbf{q}_N(x_2)})/\hat{\mu}(F)) - h(T_2, \bar{P}) \right| < \frac{\varepsilon\alpha}{16}$. (7.14)

We also require that $N > (1/\alpha)\log_2(16/\varepsilon)$, and $N > 32(N_0 + 2)/\varepsilon$.

Apply the Rohlin lemma (Theorem 3.10) to construct a tower in $\bar{X}_2$ of height $N + N_0 + 2$ with base set $\bar{F}$, covering all but $\varepsilon/32$ of X_2. For now we work only on the first N levels of the tower, $\bar{F}, T_2(\bar{F}), \ldots, T_2^{N-1}(\bar{F})$. This covers all but $\varepsilon/16$ of X_2.

Let $F = X_1 \times \bar{F}$, a base in $X_1 \times X_2$. Let $F_0 \subseteq F$ consist of those (x_1, x_2) with

(1) the $\bar{P} \times \bar{Q}, N$-name of (x_1, x_2) is $\varepsilon\alpha/16$-generic for $\hat{\mu}$; and (7.15)

(2) (a) $\left| -\frac{1}{N} \log_2(\hat{\mu}(F_{\mathbf{p}_N(x_1)})/\hat{\mu}(F)) - h(T_1, \bar{P}) \right| < \frac{\varepsilon\alpha}{16}$; and

(b) $\left| -\frac{1}{N} \log_2(\hat{\mu}(F_{\mathbf{p}_N(x_2)})/\hat{\mu}(F)) - h(T_2, \bar{P}) \right| < \frac{\varepsilon\alpha}{16}$. (7.16)

We know $\hat{\mu}(F_0) > 1 - (3\varepsilon\alpha/16)\hat{\mu}(F)$. Let

$$J = \{\mathbf{q} = \mathbf{q}_N(x_2)\text{: there is an } x_1 \text{ with } (x_1, x_2) \in F_0\},$$

and

$$I = \{\mathbf{p} = \mathbf{p}_N(x_2) : \text{there is an } x_2 \text{ with } (x_1, x_2) \in F_0\}.$$

To any element $\mathbf{q} \in J$ we can associate $I(\mathbf{Q}) \subseteq I$ where

$$I(\mathbf{q}) = \{\mathbf{p} \in J : \text{for some } (x_1, x_2) \in F_0,\ \mathbf{p}_n(x_1) = \mathbf{p} \text{ and } \mathbf{q}_N(x_2) = q\}. \quad (7.17)$$

We want to paint the tower with base $\bar{F}_{\mathbf{q}}$ with a name $\mathbf{p}$ from $I(\mathbf{q})$. This will make the pair $(\mathbf{p}, \mathbf{q})$ $\varepsilon\alpha/16$-generic for $\hat{\mu}$. To obtain (2) of (7.12), though we want, to as great a degree as possible, $\mathbf{p}$ to be unique to $\mathbf{q}$.

To be more specific, consider $\Phi = \{\varphi; J_0 \to I; J_0 \subseteq J, \varphi \text{ is } 1\text{–}1 \text{ and } \varphi(\mathbf{q}) \in I(\mathbf{q})\}$. We partially order Φ by $\varphi_1 \prec \varphi_0$ if $\mathrm{Dom}(\varphi_1) \subseteq \mathrm{Dom}(\varphi_0)$. Let φ be a maximal element of Φ. Let us compute

$$\hat{\mu}(\{(x_1, x_2) \in F; \mathbf{q}_N(x_2) \in \mathrm{dom}(\varphi)\}).$$

Notice that for any $\mathbf{q} \in J$ and $\mathbf{p} \in I$

$$\frac{\hat{\mu}(F_{\mathbf{p}})}{\hat{\mu}(F_{\mathbf{q}})} < 2^{-\alpha(1-\bar{\varepsilon}')N} < 2^{-7/8\alpha N} < \frac{\varepsilon}{16}. \quad (7.18)$$

Thus

$$\hat{\mu}(\{(x_1, x_2) : \mathbf{p}_N(x_1) \in \mathrm{Range}(\varphi)\}) < \frac{\varepsilon}{16}\hat{\mu}(\{(x_1, x_2); \mathbf{q}_N(x_2) \in \mathrm{dom}(\varphi)\})$$

$$< \frac{\varepsilon}{16}\hat{\mu}(F).$$

Since φ is maximal, $\mathrm{Range}(\varphi) \supseteq \bigcup_{\mathbf{q} \notin \mathrm{Dom}(\varphi)} I(\mathbf{q})$ as otherwise φ could be extended to some $\mathbf{q} \notin \mathrm{Dom}(\varphi)$ by assigning $\varphi(\mathbf{q}) = \mathbf{p} \in I(\mathbf{q})|_{\mathrm{Range}(\varphi)}$. Thus

$$\begin{aligned}
&\hat{\mu}(\{(x_1, x_2) \in F; \mathbf{q}_N(x_2) \in \mathrm{Dom}(\varphi)\}) \\
&\quad \geq \hat{\mu}(F_0) - \hat{\mu}(\{(x_1, x_2) \in F_0 : \mathbf{q}_N(x_2) \notin \mathrm{Dom}(\varphi)\}) \\
&\quad \geq \hat{\mu}(F_0) - \hat{\mu}(\{(x_1, x_2) \in F_0 : \mathbf{p}_N(x_1) \in \mathrm{Range}(\varphi)\}) \\
&\quad \geq \hat{\mu}(F_0) - \frac{\bar{\varepsilon}}{16}\hat{\mu}(F) \\
&\quad > \hat{\mu}(F)\left(1 - \frac{\varepsilon}{4}\right).
\end{aligned} \quad (7.19)$$

Now for $\mathbf{q} \in \mathrm{Dom}(\varphi)$ paint the name $\varphi(\mathbf{q})$ on the tower $\bar{F}_{\mathbf{q}}$, $T_2(\bar{F}_{\mathbf{q}})$, $\ldots$, $T_2^{N-1}(\bar{F}_{\mathbf{q}})$. For $x \in F_{\mathbf{q}}$, $\mathbf{q} \notin \mathrm{dom}(\varphi)$, paint $\tilde{P}(x)$ arbitrarily. Remember we have $N_0 + 2$ further levels $T^N(F)$, $\ldots$, $T^{N+N_0+1}(F)$ to paint. Paint these with the name p_2 followed by $N_0 p_1$'s followed by a p_2. This name never occurs as a name $\varphi(\mathbf{q})$ as N_0 consecutive p_1's never occur in a $T_1,\bar{P}$-name. On the remainder of X_2, let $\tilde{P}$ be arbitrary. This constructs the partition $\tilde{P}$ of X_2.

To verify (1) of (7.12), notice that by (7.19) the towers over $\bar{F}_{\mathbf{q}}$, $\mathbf{q} \in \mathrm{Dom}(\varphi)$

cover all but $(5\varepsilon/16)$ of X_2. We can regard each such tower as painted by a double name $(\varphi(\mathbf{q}), \mathbf{q})$ which we know to be $\varepsilon\alpha/16$-generic for $\hat{\mu}$. By Theorem 7.4,

$$\|(\bar{X}_1 \times \bar{X}_2, \hat{\mu}), \bar{P} \times \bar{Q}; \bar{X}_2, \tilde{P} \times \bar{Q}\| < \frac{3\varepsilon}{8} < \varepsilon.$$

For (2) of (7.12), we show how to approximate $\bar{Q}$ in $\bigvee_{i=-\infty}^{\infty} T_2^{-i}(\tilde{P})$. For a point $x \in X_2$ examine the name

$$\tilde{P}(x), \tilde{P}(T_2(x)), \ldots, \tilde{P}(T_2^{N+N_0+1}(x))$$

for an occurrence of the name $p_2, p_1, \ldots, p_1, p_2$. If $x \in T_2^j(\bar{F})$, $0 \lneq j < N$, this will occur at precisely

$$\tilde{P}(T_2^{N-j}(x)), \ldots, \tilde{P}(T_2^{N+N_0+1-j}(x)). \tag{7.20}$$

The name

$$\mathbf{p}(j, x) = \tilde{P}(T_2^{-j}(x)), \ldots, \tilde{P}(T_2^{N-1-j}(x))$$

will be in range (φ) and $\bar{Q}(x) = \varphi^{-1}(\mathbf{p}(j, x))_j$.

Thus in $\bigvee_{i=-N+1}^{N+N_0+1} T_2^{-i}(\tilde{P})$ is a partition $\tilde{Q}$ which agrees with $\bar{Q}$ on $\bigcup_{\mathbf{q} \in \mathrm{Dom}(\varphi)} \bigcup_{i=0} T_2^i(F_q)$, i.e., $\mu_2(\bar{Q} \,\Delta\, \tilde{Q}) < (5\varepsilon/16) < \varepsilon$. ■

Lemma 7.18 (Ornstein's fundamental lemma, second form) *Suppose* $(\bar{X}_1, \bar{P}_1)$ *and* $(\bar{X}_2, \bar{Q})$ *are non-periodic ergodic processes and suppose* $h(T_1, \bar{P}) > h(T_2)$. *Further, suppose* $\hat{\mu} \in J(\bar{X}_1, \bar{X}_2)$ *is an ergodic joining and* $\varepsilon > 0$. *There is, then, a non-periodic ergodic process* $(\bar{X}_3, \bar{P}_3)$ *with* $h(T_3, \bar{P}_3) > h(T_2)$. *Further, there is an ergodic joining* $\hat{\mu}_1 \in J(\bar{X}_3, \bar{X}_2)$ *so that*

(1) $\|(\bar{X}_1 \times \bar{X}_2, \hat{\mu}), \bar{P}_1 \times \bar{Q}; (\bar{X}_3 \times \bar{X}_2, \hat{\mu}_1), \bar{P}_3 \times \bar{Q}\| < \varepsilon$,

(2) $\bar{P}_3 \times X_2 \underset{\hat{\mu}_1}{\overset{\varepsilon}{\subset}} X_3 \times \mathscr{F}_2$, *and*

(3) $X_3 \times \bar{Q} \underset{\hat{\mu}_1}{\overset{\varepsilon}{\subset}} \mathscr{F}_3 \times X_2$. (7.21)

Proof First notice how this result differs from the first form. Here we suppose $h(T_1, \bar{P}_1) > h(T_2)$, not just greater than $h(T_2, \bar{Q})$. In Theorem 7.17, we constructed $\tilde{P}$ inside the $\bar{X}_2$ process. Here we only get $(\bar{X}_3, \bar{P}_3)$ joined to $(\bar{X}_2, \bar{Q})$ with approximate containment. The critical new fact is

$$h(T_3, \bar{P}_3) > h(T_2).$$

We will gain this entropy using Theorem 7.12.

Notice that

$$h(S) < h(T, P) \leq \log_2(n).$$

Let $\log_2(n) - h(S) > \alpha > 0$, $\alpha \le 1$. Also notice that if we obtain (7.21) for a refinement of $\bar{Q}$, we automatically get it for $\bar{Q}$ itself.

Select $\bar{\varepsilon}$ so that

$$H(\bar{\varepsilon}) + \bar{\varepsilon}\log_2(m) < \frac{\varepsilon}{2} \quad \text{and} \quad \bar{\varepsilon} < \frac{\varepsilon\alpha}{16} \le \frac{\varepsilon}{16}. \tag{7.22}$$

By our remark above we can assume, without loss of generality, that

$$0 \le h(T_2) - h(T_2, \bar{Q}) \le \frac{\bar{\varepsilon}\alpha}{8}. \tag{7.23}$$

Use Theorem 7.17 with error $\bar{\varepsilon}/2$. This gives us a partition $\tilde{P}$ of X_2 so that

(1) $\|(\bar{X}_1 \times \bar{X}_2, \hat{\mu}), \bar{P} \times \bar{Q}; \bar{X}_2, \tilde{P} \vee \bar{Q}\| < \dfrac{\varepsilon}{2}$, and

(2) $\bar{Q} \overset{\bar{\varepsilon}/2}{\underset{\mu_2}{\subset}} \bigvee_{i=-\infty}^{\infty} T_2^{-i}(\tilde{P}).$ (7.24)

Let $\nu_1 = \nu_{(\bar{X}_2, \tilde{P})} \in \eta_P$. If we apply Theorem 7.12, with error $\varepsilon/4$, for any N, there is a $\nu_2 \in \eta_P$ with

(1) $h_{\nu_2}(S) \ge h(T_2, \tilde{P}) + \dfrac{\varepsilon}{4}(\log_2(n) - h(T_2, \tilde{P})) \ge h(T_2, \tilde{P}) + \dfrac{\bar{\varepsilon}\alpha}{4} > h(S) + \dfrac{\varepsilon\alpha}{8}$; and

(2) $\bar{d}^N(\nu_1, \nu_2) < \dfrac{\bar{\varepsilon}}{2}.$ (7.25)

As ν_1 is ergodic, we may assume ν_2 is also.

How do we choose the parameter N? We ask two things of N, first $N > \log_2(4/\varepsilon)$, and second, as $\bar{Q} \subset_{\mu_2}^{-\bar{\varepsilon}/2} \bigvee_{i=-\infty}^{\infty} T_2^{-i}(\tilde{P})$, we ask that $\bar{Q} \subset_{\mu_2}^{\bar{\varepsilon}} \bigvee_{i=-N}^{N} T_2^{-i}(\tilde{P})$. With this choice for N, let $(\bar{X}_3, \bar{P}_3) = ((\bar{Y}_P, \nu_2), P)$ and $\hat{\mu}_1$ be an ergodic joining of $(\bar{X}_3, \bar{P}_3)$ and $(\bar{X}_2, \tilde{P})$ that achieves the $\bar{d}^N$-distance.

Notice that as (7.25) tells us

$$\bar{d}^N(\bar{X}_2, \tilde{P} \vee \bar{Q}; (\bar{X}_3 \times \bar{X}_2, \hat{\mu}_1), \bar{P} \times \bar{Q}) < \frac{\bar{\varepsilon}}{2} < \frac{\varepsilon}{2},$$

and $n > \log_2(4/\varepsilon)$, Exercise 7.7 tells us

$$\|\bar{X}_2, \tilde{P} \vee \bar{Q}; (\bar{X}_3 \times \bar{X}_2, \hat{\mu}_1), \bar{P} \times \bar{Q}\| < \varepsilon \tag{7.26}$$

and we obtain (1) of (7.21).

To see (2) of (7.21), since $P \subseteq \mathscr{F}_2$ and $\hat{\mu}_1(\bar{P}_3 \times X_2 \Delta X_3 \times \tilde{P})^- < (\bar{\varepsilon}/2) < \varepsilon$, we have $\bar{P}_3 \times X_2 \subset_{\hat{\mu}_1}^{\varepsilon} X_3 \times \mathscr{F}_2$.

Conclusion (3) of (7.21) is slightly more delicate. We know

$$Q \overset{\bar{\varepsilon}}{\underset{\mu_2}{\subset}} \bigvee_{i=-N}^{N} T_2^{-i}(\tilde{P}),$$

i.e., there is a partition $\bar{Q}_0 \subseteq \bigvee_{i=-N}^{N} T_2^{-i}(\tilde{P})$ with $\mu_2(\bar{Q}\,\Delta\,\bar{Q}_0) < \bar{\varepsilon}$. As

$$\hat{\mu}_1\left(\bigvee_{i=-N}^{N} T_3^{-i}(\bar{P}_3) \times X_2\,\Delta\,X_3 \times \bigvee_{i=-N}^{N} T_2^{-i}(\tilde{P})\right) < \frac{\bar{\varepsilon}}{2},$$

in $\bigvee_{i=-N}^{N} T_3^{-i}(\bar{P}_3)$, we can define $\bar{Q}_1$ to be the union of precisely the same P-names as form $\bar{Q}_0$ in $\bigvee_{i=-N}^{N} T_2^{-i}(\tilde{P})$. This gives $\hat{\mu}_1(\bar{Q}_1 \times X_2\,\Delta\,X_3 \times \bar{Q}_0) < (\bar{\varepsilon}/2)$ and so $X_3 \times \bar{Q} \subset_{\hat{\mu}_1}^{3\bar{\varepsilon}/2} \mathscr{F}_3 \times X_2$. ■

7.5 Krieger's finite generator theorem

Theorem 7.19 (Krieger's finite generator theorem (Krieger 1970)) *Suppose $\bar{X}$ is ergodic and $h(T) < \log_2(n)$. There is then a partition $\bar{P}$ of X into n sets with $\mathscr{F} = \bigvee_{i=-N}^{N} T^{-i}(\bar{P})$.*

Before we prove this, some remarks are in order. First, notice the entropy bound $h(T) < \log_2(n)$ is necessary as the only measure $\nu \in \eta_P$ of entropy $\log_2(n)$ is the full n-shift, which at the very least is a K-system. If $\bar{X}$ has entropy $\log_2(n)$ but is *not* a K-system, then certainly it cannot have an n-set generator.

Second, our proof is not Krieger's original, but is due to Burton and Rothstein who first noticed the deep connection between the Ornstein and Krieger theorems. This connection is so natural that the fundamental lemma designed to complete Ornstein's theorem is in fact more easily applied to prove the Krieger result.

Notice that by Theorem 6.8 the conclusion of Krieger's theorem is equivalent to saying there is a joining $\mu \in J((Y_P, \nu), X)$ so that

(1) $P \times X \subset_{\hat{\mu}} Y_P \times \mathscr{F}$; and

(2) $Y_P \times \mathscr{F} \subset_{\mu} \mathscr{B} \times X$. (7.27)

Letting $\{Q_i\}$ be a refining and generating tree of partitions in X, with $T(Q_i) \vee T^{-i}(Q_i) \subseteq Q_{i+1}$, (2) of (7.27) is equivalent to

$$Y_P \times Q_i \subset_{\hat{\mu}} \mathscr{B} \times X. \tag{7.28}$$

Conditions (1) and (7.28) look suspiciously similar to the conclusion of Lemma 7.18.

Our first step is to establish the space in which these joinings sit. Let $\tilde{J}$ consist of all joinings of $(\bar{Y}_P, \nu)$ and $\bar{X}$, where $\nu \in \eta_P$. This space is half-way between a purely symbolic space like η_P and a joining space. The first coordinate is symbolic, and its marginal measure arbitrary. The second coordinate is fixed. There is a natural metric topology on $\tilde{J}$, again a weak* topology given by

$$\|\hat{\mu}_1, \hat{\mu}_2\| = \sum_{n=1}^{\infty} \frac{1}{2^n} \|\hat{\mu}_1(P_n \times Q_n), \hat{\mu}_2(P_n \times Q_n)\|_1 \tag{7.29}$$

where

$$\|\hat{\mu}_1(P_n \times Q_n), \hat{\mu}_2(P_n \times Q_n)\|_1 = \frac{1}{2} \sum_{A \times B \in P_n \times Q_n} |\hat{\mu}_1(A \times B) - \hat{\mu}_2(A \times B)| \leq 1. \tag{7.30}$$

Lemma 7.20 *The metric space $(\bar{J}, \|\cdot, \cdot\|)$ is compact and convex.*

Proof Convexity is obvious. For compactness, notice that $H_n = P_n \times Q_n$ form a generating tree of partitions. Any additive set function $\hat{\mu}_0$ on this tree which agrees with μ on its second marginal will extend to a measure in $\tilde{J}$. This is because the only possible empty chains in the tree must be empty in the Q_n tree. The additive set functions are clearly $\|\cdot, \cdot\|$ compact. ∎

Definition 7.7 Let $\hat{J}$ be the weak* closure of those $\hat{\mu} \in \tilde{J}$ with

(1) $\hat{\mu}$ ergodic; and

(2) $h_\nu(S) > h(T)$ for ν the first marginal of $\hat{\mu}$.

Corollary 7.21 *$\hat{J}$ is a non-empty, compact space.*

Proof Let $(\bar{Y}_P, \nu_0)$ be the full n-shift, $(n^{-1}, n^{-1}, \ldots, n^{-1})$, and $\hat{\mu} = \nu_0 \times \mu$. As $(\bar{Y}_P, \nu_0)$ is weakly mixing, $\hat{\mu}$ is ergodic. As $h_\nu(S) = \log_2(n) > h(T)$, $\hat{\mu} \in \hat{J}$. ∎

It is a very delicate task to identify $\hat{J}$ more precisely. It is not necessarily all elements of $\tilde{J}$ whose first coordinate $\nu \in \eta_P$ has $h_\nu(S) \geq h(T)$, although it is contained in this set. The ergodic elements of $\tilde{J}$ may not be dense in it. The precise nature of $\hat{J}$ depends delicately on $\bar{X}$.

What we will show is that the $\hat{\mu} \in \hat{J}$ satisfying our earlier conditions (7.27) and (7.28) are a dense G_δ. Let

$$\mathcal{O}(n) = \left\{ \hat{\mu} \in \bar{J} : P \times X \overset{1/n}{\underset{\hat{\mu}}{\subset}} Y_P \times \mathscr{F} \text{ and } Y_P \times Q_n \overset{1/n}{\underset{\hat{\mu}}{\subset}} \mathscr{B} \times X \right\}.$$

It is easy to see that any $\hat{\mu} \in \bigcap_n \mathcal{O}_n$ will satisfy (7.27) and (7.28).

Proof of Theorem 7.19 What we will show is that the sets $\mathcal{O}(n)$ are open and dense in $\hat{J}$. The Baire category theorem finishes the result, telling us $\bigcap_n \mathcal{O}(n)$ is a dense G_δ in $\hat{J}$.

To see that $\mathcal{O}(n)$ is dense, notice that if $\hat{\mu} = \mathcal{O}(n)$, then for some N large enough there are partitions $\bar{P}' \subseteq Q_N$ and $\bar{Q}'_n \subseteq P_N$ with

$$|\hat{\mu}(Y_P \times \bar{P}' \Delta P \times X)| < \varepsilon$$

and

$$|\hat{\mu}(\bar{Q}'_n \times X \,\Delta\, Y_P \times Q_n)| < \varepsilon.$$

These are strict inequalities and all the sets in these expressions are finite unions of elements of our tree. Thus for some $\delta > 0$, if $\|\hat{\mu}', \hat{\mu}\| < \delta$, these inequalities still hold with $\hat{\mu}$ replaced by $\hat{\mu}'$. Hence $\hat{\mu}' \in \mathcal{O}(n)$ and it is open.

To show $\mathcal{O}(n)$ is dense, choose $\hat{\mu} \in \hat{J}$. We can assume $\hat{\mu}$ is ergodic and $h_\nu(S) > h(T)$ as such μ *are* dense in $\hat{J}$. Apply Lemma 7.18 with $\bar{P} = P, \bar{Q} = Q_n$, $\hat{\mu}$ as given, and any $\varepsilon \leq 1/n$. We get a $\hat{\mu}_1$ with

(1) $\|\hat{\mu}, \hat{\mu}_1\| < \varepsilon$;

(2) $P \times X \overset{1/n}{\underset{\hat{\mu}_1}{\subset}} Y_P \times \mathscr{F}$, and

(3) $Y_P \times Q_n \overset{1/n}{\underset{\hat{\mu}_1}{\subset}} \mathscr{B} \times X.$ (7.31)

As $\hat{\mu}_1$ is ergodic, and its first marginal ν_1 satisfies $h_{\nu_1}(S) > h(T)$, $\hat{\mu}_1 \in \mathcal{O}(n)$. Hence $\mathcal{O}(n)$ is dense in $\bar{J}$. ■

Corollary 7.22 *For any $\bar{X}$ with $h(T) < \log(n)$ there is a measure $\nu \in \eta_P$ with $(\bar{Y}_P, \nu)$ isomorphic to $\bar{X}$.* ■

Exercise 7.9 Suppose $\bar{X}$ has entropy less than $\log(n)$ and for $\nu \in \eta_P$, $h_\nu(S) \geq h(T)$. Given any $\varepsilon > 0$, show that there is a generating partition $\bar{P}$ of $\bar{X}$ with

$$\|\bar{X}, \bar{P}; (\bar{Y}_P, \nu), P\| < \varepsilon.$$

Note: ν is *not* assumed ergodic. This shows that in

$$\{\nu \in \eta_P; h_\nu(S) \geq h(T)\},$$

the set of measures isomorphic to $\bar{X}$ are dense. They of course cannot be a G_δ, as any two such isomorphism classes are either disjoint or equal.

7.6 Ornstein's isomorphism theorem

Ornstein's theorem is much more than just that two Bernoulli shifts of equal entropy are isomorphic. It, in fact, identifies a certain property which allows for the proof of the isomorphism theorem. Bernoulli shifts happen to satisfy it. This property, and a rather long list of derivative properties, characterize those processes isomorphic to Bernoulli shifts.

Our intention here is to prove the isomorphism theorem. Hence we will not delve too deeply into the world of such processes. The bibliography will direct the interested reader to sources of this material. We begin with its definition.

Definition 7.8 We say an ergodic process $(\bar{X}, \bar{P})$ is *finitely determined* if for any $\varepsilon > 0$, there is a $\delta > 0$ so that if $(\bar{X}_1, \bar{P}_1)$ is ergodic and

(1) $\|\bar{X}, \bar{P}; \bar{X}_1, \bar{P}_1\| < \delta$; and

(2) $h(T_1, P_1) > h(T, P) - \delta$; (7.32)

then

(3) $\bar{d}(\bar{X}, \bar{P}; \bar{X}_1, \bar{P}_1) < \varepsilon$. (7.33)

To complete the isomorphism theorem we will show first that any two finitely determined processes of equal entropy are isomorphic, and second that Bernoulli shifts are finitely determined. We will show a little more. In fact, we will see all mixing Markov chains are finitely determined, hence any two such of equal entropy are isomorphic. First a small technical fact.

Lemma 7.23 *If $(\bar{X}, \bar{P})$ is finitely determined, and non-trivial, then $h(T, \bar{P}) > 0$.*

Proof By non-trivial we mean $\bar{P}$ consists of more than one set of positive measure. Suppose $h(T, \bar{P}) = 0$. For any $0 < \delta < \frac{1}{2}\sum \mu(p_i)^2 = a$, use Theorem 7.5 to find $(\bar{X}_2, \bar{P}_2)$, a K-system with

(1) $\|\bar{X}, \bar{P}; \bar{X}_1, \bar{P}_1\| < \delta$; and

(2) $h(T_2, P_2) < \delta$.

$\bar{X}_1$ and $\bar{X}_2$ are disjoint, so

$$\begin{aligned}\bar{d}(\bar{X}, \bar{P}; \bar{X}_1, \bar{P}_1) &= \sum_{i=1}^{n} \mu(\bar{p}_i)\mu_2(\bar{p}_i^2)\\ &\geq \sum_{i=1}^{n} \mu(p_i)^2 - \delta\\ &> \frac{1}{2}\sum_{i=1}^{n} \mu(p_i)^2 = a > 0.\end{aligned}$$

Hence $(\bar{X}, \bar{P})$ cannot be finitely determined. ■

Let $(\bar{X}_1, \bar{P})$ be finitely determined, and $(\bar{X}_2, \bar{Q})$ another ergodic process with

$$h(T_1, \bar{P}) = h(T_2, \bar{Q}).$$

We assume $\bar{P}$ and $\bar{Q}$ are generators of their respective processes.

Let $\bar{J}$ be the weak* closure of the ergodic joinings in $J(\bar{X}_1, \bar{X}_2)$. As long as X_1 exists, $\bar{J}$ is not empty, as it is the closure of the extreme points of $J(\bar{X}_1, \bar{X}_2)$. In $\bar{J}$ we use the choice $(\bar{P}_n \times \bar{Q}_n)$ for the generating tree defining the weak* metric.

Theorem 7.24 *Those elements $\hat{\mu} \in \bar{J}$ with $\bar{P} \times X_2 \underset{\hat{\mu}}{\subset} X_1 \times \mathscr{F}_2$ are a dense G_δ.*

Proof We define

$$\mathcal{O}(n) = \left\{ \hat{\mu} \in \bar{J} : \bar{P} \times X_2 \overset{1/n}{\underset{\hat{\mu}}{\subset}} X_1 \times \mathscr{F}_2 \right\}. \tag{7.34}$$

The proof that $\mathcal{O}(n)$ is open follows the same lines as that part of Theorem 7.19.

To show denseness, we work as follows. Let $\hat{\mu} \in \bar{J}$, and $1/2n > \varepsilon > 0$ be given. We want to find $\hat{\mu}_1 \in \mathcal{O}(n)$ with $\| \hat{\mu}, \hat{\mu}_1 \| < \varepsilon$. We may, without loss of generality, assume $\hat{\mu}$ is ergodic.

We know by Theorem 7.7 that if

$$\bar{d}((\bar{X}_1 \times \bar{X}_2, \tilde{\mu}_1), \bar{P} \times \bar{Q}; (\bar{X}_1 \times \bar{X}_2, \tilde{\mu}_2), \bar{P} \times \bar{Q}) < \bar{\varepsilon} = \frac{\varepsilon}{4 \log(2/\varepsilon) + 4}$$

then $\|(\bar{X}_1 \times \bar{X}_2, \tilde{\mu}_1), \bar{P} \times \bar{Q}; (\bar{X}_1 \times \bar{X}_2, \tilde{\mu}_2), \bar{P} \times \bar{Q}\| < \varepsilon$. Using $\bar{\varepsilon}/4$ in the definition of finitely determined for $(\bar{X}_1, P)$, we obtain a $\delta > 0$. We assume $\delta < \bar{\varepsilon}/4 < \varepsilon/4$.

As a first step, combining Lemma 5.10 and Exercise 7.8, we can find a partition $\bar{Q}_1$ of X_2 with

$$\mu_2(\bar{Q} \,\Delta\, \bar{Q}_1) < \frac{\bar{\varepsilon}}{4}$$

and

$$h(T_2, \bar{Q}) - \delta < h(T_2, \bar{Q}_1) < h(T_2, \bar{Q}).$$

Now

$$h(T_1, \bar{P}) > h(T_2, \bar{Q}_1)$$

and $\hat{\mu}$ can be regarded as an ergodic joining of $(\bar{X}_1, \bar{P})$ and $(\bar{X}_2, \bar{Q}_1)$. By Lemma 7.18 there is an ergodic process $(\bar{X}_3, \bar{P}_3)$ and an ergodic joining $\hat{\mu}_0 \in J(\bar{X}_3, \bar{X}_2)$ with

(1) $\|(\bar{X}_1 \times \bar{X}_2, \tilde{\mu}), \bar{P} \times \bar{Q}_1; (\bar{X}_3 \times \bar{X}_2, \tilde{\mu}_1), \bar{P}_3 \times \bar{Q}_1\| < \delta$; and

(2) $\bar{P}_3 \times X_2 \overset{1/2n}{\underset{\hat{\mu}_0}{\subset}} X_1 \times \bigvee_{i=-\infty}^{\infty} T_2^{-i}(\bar{Q}_1)$; most importantly,

(3) $h(T_3, \bar{P}_3) > h(T_2, \bar{Q}_1) > h(T_1, \bar{P}) - \delta.$ (7.35)

By (1) of (7.35), (1) $\|\bar{X}_1 \times \bar{P}; \bar{X}_3 \times \bar{P}_3\| < \delta$ and by (3), (2) $h(T_3, \bar{P}_3) > h(T, \bar{P}) - \delta$.

By our choice of δ, there is an ergodic $\tilde{\mu} \in J(\bar{X}_1, \bar{X}_3)$ with

$$\mu(\bar{P} \times \bar{X}_3 \,\Delta\, X_1 \times \bar{P}_3) < \frac{\bar{\varepsilon}}{4}.$$

Let $\bar{\mu}$ be almost any ergodic component of the relatively independent joining of $\hat{\mu}_0 \in J(\bar{X}_3, \bar{X}_2)$ and $\tilde{\mu} \in J(\bar{X}_1, \bar{X}_3)$ over their common factor $\bar{X}_3$, hence an ergodic joining of $\bar{X}_3$, $\bar{X}_2$ and $\bar{X}_1$.

Let $\hat{\mu}_1 \in \bar{J}$ be the restriction of $\bar{\mu}$ to these two components. To compute $\|(\bar{X}_1 \times \bar{X}_2, \hat{\mu}), \bar{P} \times \bar{Q}; (\bar{X}_1 \times \bar{X}_2, \hat{\mu}_1), \bar{P} \times \bar{Q}\|$, notice it is at most

$$
\begin{aligned}
&(1)\ \ \|(\bar{X}_1 \times \bar{X}_2, \hat{\mu}), \bar{P} \times \bar{Q}; (\bar{X}_1 \times \bar{X}_2, \hat{\mu}), \bar{P} \times \bar{Q}_1\| + \\
&(2)\ \ \|(\bar{X}_1 \times \bar{X}_2, \hat{\mu}), \bar{P} \times \bar{Q}_1; (\bar{X}_3 \times \bar{X}_2, \hat{\mu}_0), \bar{P}_3 \times \bar{Q}_1\| + \\
&(3)\ \ \|(\bar{X}_3 \times \bar{X}_2, \hat{\mu}_0), \bar{P}_3 \times \bar{Q}_1; (\bar{X}_1 \times \bar{X}_2, \hat{\mu}_1), \bar{P} \times \bar{Q}_1\| + \\
&(4)\ \ \|(\bar{X}_1 \times \bar{X}_2, \hat{\mu}_1), \bar{P} \times \bar{Q}_1; (\bar{X}_1 \times \bar{X}_2, \hat{\mu}_1), \bar{P} \times \bar{Q}\|.
\end{aligned}
\tag{7.36}
$$

The pairs of processes in terms (1), (2), and (4) of (7.36) are within a $\bar{d}$-distance $\bar{\varepsilon}/4$. Hence each of these three is less than or equal to $\varepsilon/4$. Term (3) we already know is not greater than $\varepsilon/4$. Hence $\hat{\mu}$ and $\hat{\mu}_1$ are weak* less than ε apart.

We know

$$\bar{P}_3 \times X_1 \underset{\hat{\mu}_0}{\overset{1/2n}{\subset}} X_3 \times \mathscr{F}_2,$$

and

$$\bar{\mu}(\bar{P}_3 \times X_2 \times X_1 \,\Delta\, X_3 \times X_2 \times \bar{P}) = \tilde{\mu}(\bar{P} \times X_3 \,\Delta\, X_1 \times \bar{P}_3) < \frac{\bar{\varepsilon}}{4} < \frac{1}{2n}.$$

Thus

$$\bar{P} \times X_2 \underset{\hat{\mu}_0}{\overset{1/n}{\subset}} X_1 \times \mathscr{F}_2,$$

and $\hat{\mu}_1 \in \mathcal{O}(n)$. ∎

Before going on to the almost obvious Corollary 7.25, we stop to make some remarks. What Theorem 7.24 tells us is that a finitely determined process can be embedded as a factor in any process of equal entropy. One easily concludes that it can also be embedded in any process of greater entropy, as such always have factors of any smaller entropy. Restricted to the case of Bernoulli shifts, this says a Bernoulli shift can be embedded as a factor of any system with equal or greater entropy. This deep fact is originally due to Sinai.

As with Krieger's theorem, we have not precisely identified $\bar{J}$. Here, in fact, it is all of $J(\bar{X}_1, \bar{X}_2)$. Once we know $\bar{X}_1$ is isomorphic to a Bernoulli shift, it is relatively easy to show the ergodic joinings are dense in $J(\bar{X}_1, \bar{X}_2)$.

Corollary 7.25 (Ornstein's isomorphism theorem, one form) *Suppose both $(\bar{X}_1, P)$ and $(\bar{X}_2, \bar{Q})$ are finitely determined, with $\bar{P}$ and $\bar{Q}$ generators, and $h(T_1) = h(T_2)$. Those elements of $\bar{J}$ supported on graphs of isomorphisms are a dense G_δ in $\bar{J}$. Hence the two systems are isomorphic.*

Proof By Theorem 7.24, those $\hat{\mu}$ with both

$$\bar{P} \times X_2 \underset{\hat{\mu}}{\subset} X_1 \times \mathscr{F}_2$$

and

$$X_1 \times \bar{Q} \underset{\hat{\mu}}{\subset} \mathscr{F}_1 \times X_2$$

are a dense G_δ. As $\bar{P}$ and $\bar{Q}$ generate,

$$\mathscr{F}_1 \times X_2 \underset{\hat{\mu}}{=} X_1 \times \mathscr{F}_2.$$

and Theorem 6.8 completes the result. ∎

Exercise 7.10 Suppose $(\bar{X}, \bar{P})$ is finitely determined and $\bar{Q}$ is another generating partition. Show that $(\bar{X}, \bar{Q})$ is also finitely determined. Thus finitely determined is a property of $\bar{X}$. Hint: Using finite code approximations, show that any $(\bar{X}_1, \bar{Q}_1)$ close to $(\bar{X}, \bar{Q})$ in entropy and distribution contains a copy of $(\bar{X}_1, \bar{P}_1)$ close to $(\bar{X}, \bar{P})$ in entropy and distribution. Use the finite codes to bring the $\bar{d}$-closeness of $(\bar{X}, \bar{P})$ and $(\bar{X}_1, \bar{P}_1)$ back to $(\bar{X}, \bar{Q})$ and $(\bar{X}_1, \bar{Q}_1)$.

In fact, any partition of a finitely determined process is finitely determined. This deep result of Ornstein and Weiss can be found in Ornstein (1974). Thus all factor algebras of finitely determined systems are themselves finitely determined.

7.7 Weakly Bernoulli processes

To complete our picture of the isomorphism theorem we want to verify that mixing Markov chains are finitely determined. We begin with a property of mixing Markov chains.

Definition 7.9 We say a process $(\bar{X}, \bar{P})$ is *weakly Bernoulli*, if for any $\varepsilon > 0$ there is a $k > 0$ so that for all N

$$\left\| D\left(\bigvee_{i=k}^{k+N-1} T^{-i}(\bar{P}) | \mathscr{P}_{\bar{P}}\right) - D\left(\bigvee_{i=k}^{k+N-1} T^{-i}(P)\right)\right\|_1 < \varepsilon,$$

i.e.,

$$\tfrac{1}{2} \sum_{B \in \bigvee_{i=k}^{k+N-1} T^{-i}(\bar{P})} \int |E(B|\mathscr{P}_{\bar{P}})(x) - \mu(B)| \, \mathrm{d}\mu(x) < \varepsilon.$$

Lemma 7.26 *Mixing Markov chains are weakly Bernoulli.*

Proof If $(\bar{X}, \bar{P})$ is Markov, then for all $k > 0$ and N,

$$\left\| D\left(\bigvee_{i=k}^{k+N-1} T^{-i}(\bar{P})|\mathscr{P}_{\bar{P}}\right) - D\left(\bigvee_{i=k}^{k+N-1} T^{-i}(\bar{P})\right)\right\|_1$$

$$= \|D(T^{-k}(\bar{P})|\mathscr{P}_{\bar{P}}) - D(T^{-k}(\bar{P}))\|_1.$$

As a mixing Markov chain is a K-system, Proposition 5.32 finishes the result.

■

Although the weakly Bernoulli property is known to be strictly weaker than finitely determined, it is the property most often applied. Hyperbolic toral automorphisms, for example, are proven Bernoulli by showing they are weakly Bernoulli for an appropriately chosen partition Ornstein (1974).

We have seen earlier (Corollary 5.26) that entropy is convex. We need a uniformity in this to proceed.

Lemma 7.27 *For n fixed let $\Pi^n = \{(\pi_1, \dots, \pi_n); \pi_i \geq 0, \sum \pi_i = 1\}$, the space of probability n-vectors. Suppose ν is a Borel probability measure on Π^n. Given $\varepsilon > 0$ there is a $\delta = \delta(\varepsilon, n) > 0$ so that if $H(\int \pi \, d\nu) - \int H(\pi) \, d\nu < \delta$, then*

$$\frac{1}{2}\int \left|\pi - \int \pi \, d\nu\right| d\nu < \varepsilon.$$

Proof We know from Corollary 5.26 that

$$\int H(\pi) \, d\nu \leq H\left(\int \pi \, d\nu\right),$$

equality holding iff ν is a point mass at $\int \pi \, d\nu$, i.e.,

$$\int \left|\pi - \int \pi \, d\nu\right| d\nu = 0.$$

Now $\int H(\pi) \, d\nu$, $H(\int \pi \, d\nu)$ and $\int |\pi - \int \pi \, d\nu| \, d\nu$ are all weak* continuous functions of ν. Compactness in weak* of the Borel probability measures completes the result. ■

Smorodinsky (1971) has shown that in fact δ does not depend on n. Our non-constructive argument of this uniform convexity of entropy completely misses this.

Lemma 7.28 *Suppose $(\bar{X}_1, \bar{P}_1)$ is ergodic and for some k, N,*

$$\left\| D\left(\bigvee_{i=k}^{k+N-1} T_1^{-i}(\bar{P}_1)|\mathscr{P}_{\bar{P}_1}\right) - D\left(\bigvee_{i=k}^{k+N-1} T_1^{-i}(\bar{P}_1)\right)\right\|_1 < \varepsilon.$$

Given any $\bar{\varepsilon} > 0$ there is a δ so that if $(\bar{X}_2, \bar{P}_2)$ satisfies

(1) $\|\bar{X}_1, \bar{P}_1; \bar{X}_2, \bar{P}_2\| < \delta$; *and*

(2) $h(T_2, \bar{P}_2) > h(T_1, \bar{P}_1) - \delta$; *then* (7.37)

(3) $$\left\| D\left(\bigvee_{i=k}^{k+N-1} T_2^{-i}(\bar{P}_2)|\mathscr{P}_{\bar{P}_2}\right) - D\left(\bigvee_{i=k}^{k+N-1} T_2^{-i}(\bar{P}_2)\right)\right\|_1 < \varepsilon + \bar{\varepsilon}. \tag{7.38}$$

Proof We know that for $j = 1, 2$,

$$(N+k)h(T_j, \bar{P}_j) = \int I\left(\bigvee_{i=0}^{k+N-1} T_j^{-i}(\bar{P}_j)|\mathscr{P}_{\bar{P}_j}\right)\mathrm{d}\mu_j.$$

There are n^{k+N} elements in P^{k+N}. Choose $\delta_1 = \delta(\bar{\varepsilon}/8, n^{k+N})$ of Lemma 7.27. Choose M so that

$$0 \leq \int I\left(\bigvee_{i=0}^{k+N-1} T_1^{-i}(\bar{P}_1)\middle| \bigvee_{i=-1}^{-M} T_1^{-i}(\bar{P}_1)\right) - I\left(\bigvee_{i=0}^{k+N-1} T_1^{-i}(\bar{P}_1)|\mathscr{P}_{\bar{P}_1}\right)\mathrm{d}\mu_1 < \frac{\delta_1\bar{\varepsilon}}{8}.$$

If δ in (1) of (7.37) is small enough,

$$\left|\int I\left(\bigvee_{i=0}^{k+N-1} T_1^{-i}(\bar{P}_1)\middle| \bigvee_{i=-1}^{-M} T_1^{-i}(\bar{P}_1)\right)\mathrm{d}\mu_1 - \int I\left(\bigvee_{i=0}^{k+N-1} T_2^{-i}(\bar{P}_2)\middle| \bigvee_{i=-1}^{-M} T_2^{-i}(\bar{P}_2)\right)\mathrm{d}\mu_2\right| < \frac{\delta_1\bar{\varepsilon}}{8}$$

as these integrals depend only on μ_j restricted to $\bigvee_{i=-M}^{k+N-1} T_j^{-i}(\bar{P}_j)$, $j = 1, 2$, respectively.

If $\delta < (\delta_1\bar{\varepsilon}/4(N+k))$ in (2) of (7.37) then

$$0 \leq \int\left(I\left(\bigvee_{i=0}^{k+N-1} T_2^{-i}(\bar{P}_2)\middle| \bigvee_{i=-1}^{-M} T_2^{-i}(\bar{P}_2)\right)\right) - I\left(\bigvee_{i=0}^{k-N-1} T_2^{-i}(\bar{P}_2)|\mathscr{P}_{\bar{P}_1}\right)\bigg)\mathrm{d}\mu_2 < \frac{3\delta_1\bar{\varepsilon}}{8}.$$

Thus for a subset $G \subseteq \bigvee_{i=-1}^{-M} T^{-i}(\bar{P}_2)$, $\mu_2(G) > 3\bar{\varepsilon}/8$, for $B \subset G$, an atom of $\bigvee_{i=-1}^{-M} T^{-i}(\bar{P}_2)$,

$$H\left(D\left(\bigvee_{i=0}^{N+k-1} T_2^{-i}(\bar{P}_2)|B\right)\right) - \frac{1}{\mu_2(B)}\int_B H\left(D\left(\bigvee_{i=0}^{N+k-1} T_2^{-i}(\bar{P}_2)|\mathscr{P}_{\bar{P}_2}\right)\right)\mathrm{d}\mu_2 < \delta_1.$$

By Lemma 7.27, for such a $B \subset G$,

$$\frac{1}{2\mu_2(B)}\int_B \left|D\left(\bigvee_{i=0}^{N+k-1} T_2^{-i}(\bar{P}_2)|B\right) - D\left(\bigvee_{i=0}^{N+k-1} T_2^{-i}(\bar{P}_2)|\mathscr{P}_{\bar{P}_2}\right)\right|\mathrm{d}\mu_2 < \frac{\bar{\varepsilon}}{8}. \tag{7.39}$$

Outside G this integrand (7.39) is bounded by 2. Thus

$$\left\|D\left(\bigvee_{i=0}^{k+N-1} T_2^{-i}(\bar{P}_2)\middle| \bigvee_{i=-1}^{-M} T_2^{-i}(\bar{P}_2)\right) - D\left(\bigvee_{i=0}^{k+N-1} T_2^{-i}(\bar{P}_2)|\mathscr{P}_{\bar{P}_2}\right)\right\|_1 < \frac{\bar{\varepsilon}}{8} + \frac{3\bar{\varepsilon}}{4} = \frac{7\bar{\varepsilon}}{8}.$$

Now

$$\left\| D\left(\bigvee_{i=k}^{k+N-1} T_2^{-i}(\bar{P}_2)|\mathscr{P}_{\bar{P}_2}\right) - D\left(\bigvee_{i=k}^{k+N-1} T_2^{-i}(\bar{P}_2)\right)\right\|_1 \leq$$

$$\text{(a)} \quad \left\| D\left(\bigvee_{i=0}^{k+N-1} T_2^{-i}(\bar{P}_2)|\mathscr{P}_{\bar{P}_2}\right) - D\left(\bigvee_{i=0}^{k+N-1} T_2^{-i}(\bar{P}_2)\,\middle|\,\bigvee_{i=-1}^{-M} T_2^{-i}(\bar{P}_2)\right)\right\|_1 +$$

$$\text{(b)} \quad \left\| D\left(\bigvee_{i=0}^{k+N-1} T_2^{-i}(\bar{P}_2)\,\middle|\,\bigvee_{i=-1}^{+M} T_2^{-i}(\bar{P}_2)\right) - D\left(\bigvee_{i=0}^{k+N-1} T_2^{-i}(\bar{P}_2)\right)\right\|_1 . \quad (7.40)$$

If δ in (1) of (7.37) is small enough, (b) of (7.40) is $\leq \varepsilon + (\bar{\varepsilon}/8)$ and we are done. ■

We will use Corollary 7.11 to show that weakly Bernoulli processes are finitely determined. This involves the construction of maps φ which match names well. We will need the following technical argument.

Lemma 7.29 *Suppose $(X_1, \mathscr{F}_1, \mu_1)$ and $(X_2, \mathscr{F}_2, \mu_2)$ are non-atomic Lebesgue spaces, $\bar{A}_1$ and $\bar{A}_2$ partition X_1 and X_2, and $\varphi : X_1 \to X_2$, is a measure-preserving 1–1 map. Suppose $\bar{Q}_1$ and $\bar{Q}_2$ are further partitions of X_1 and X_2, respectively. We can then find a 1–1 measure-preserving $\varphi' : X_1 \to X_2$ so that for any $a_1 \in A_1$ and $a_2 \in A_2$,*

$$\mu_1(a_1 \cap \varphi^{-1}(a_2)) = \mu_1(a_1 \cap \varphi'^{-1}(a_2)).$$

Further,

(1) $D(\bar{Q}_1 | a_1 \cap \varphi'^{-1}(a_2)) = D(\bar{Q}_1 | a_1)$;

(2) $D(\bar{Q}_2 | \varphi'(a_1) \cap a_2) = D(\bar{Q}_2 | a_2)$; *and*

(3) $\mu_1(\{x \in a_1 \cap \varphi'^{-1}(a_2) : \bar{Q}_1(x) \neq \bar{Q}_2(\varphi'(x))\}) =$

$$\mu(a_1 \cap \varphi'^{-1}(a_2)) \, \|D(\bar{Q}_1 | a_1) - D(\bar{Q}_2 | a_2)\|_1 . \quad (7.41)$$

Proof As X_1 and X_2 are non-atomic, it is clear we can move the sets $\varphi^{-1}(A_2) \cap a_1$ without changing their mass so that (1) is satisfied and $D(\bar{Q}_2 | \varphi'(a_1) \cap a_2)$ remains $D(\bar{Q}_2 | \varphi(a_1) \cap a_2)$. Repeat this step in X_2 to obtain (2) without losing (1).

Within each set $a_1 \cap \varphi'(a_2)$ we can further perturb φ' so as to map as much of $q_i^1 \cap a_1$ to $q_i^2 \cap a_2$ as possible, i.e., a set of measure $\min(\mu_1(q_i^1 \cap a_1), \mu_2(q_i^2 \cap a_2))$. Map the rest of $a_1 \cap \varphi'^{-1}(a_2)$ to the remainder of $\varphi'(a_1) \cap a_2$ to get the real φ'. We compute

$$\mu_1(\{x \in a_1 \cap \varphi'(a_2) : \bar{Q}_1(x) \neq \bar{Q}_2(\varphi'(x))\})$$
$$= \mu_1(a_1 \cap \varphi'(a_2)) \, \|D(\bar{Q}_1 | a_1) - D(\bar{Q}_2 | a_2)\|_1 .$$

Our next result will complete the proof that weakly Bernoulli implies finitely determined. It is an inductive construction of maps φ_r enabling us to apply

Corollary 7.11. The notation becomes very complex, but essentially all we are doing is successively matching more and more of the $T,\bar{P}_1$-names to $T_2,\bar{P}_2$-names, without disturbing the statistics of our earlier matching. Our starting point is the hypothesis and conclusion of Lemma 7.28.

Theorem 7.30 *Suppose* $(\bar{X}_1, \bar{P}_1)$ *and* $(\bar{X}_2, \bar{P}_2)$ *satisfy*

(1) $\|D(\bigvee_{i=k}^{k+N-1} \bar{T}_j^{-i}(\bar{P}_j)|\mathscr{P}_{\bar{P}_j}) - D(\bigvee_{i=k}^{k+N-1} \bar{T}_j^{-i}(P_j))\|_1 < \varepsilon,$ *for* $j = 1$ *and* 2; *and*

(2) $$\left\| D\left(\bigvee_{i=k}^{k+N-1} \bar{T}_1^{-i}(\bar{P}_1)\right) - D\left(\bigvee_{i=k}^{k+N-1} \bar{T}_2^{-i}(P_2)\right)\right\|_1 < \varepsilon,$$

where N, K, ε *are* > 0. (7.42)

Then

(3) $$\bar{d}(\bar{T}_1, \bar{P}_1; \bar{T}_2, \bar{P}_2) < 3\varepsilon + \frac{k}{N+k}. \tag{7.43}$$

Proof We will use Corollary 7.11. The sets $A_r = X_1$, and the lengths $N_r = r(N + k)$.

To begin, φ_0 is any 1–1 measure-preserving map $X_1 \to X_2$. By Lemma 7.29 and (2) of (7.42) we can construct $\varphi_1 : X_1 \to X_2$ with

$$\begin{aligned}
&\int \bar{d}_{(N+k)}(\mathbf{p}^1_{(N+k)}(x), \mathbf{p}^2_{(N+k)}(\varphi_1(x))\,\mathrm{d}\mu \\
&\le \frac{k}{N+k} + \mu_1(\{x; \mathbf{p}^1_N(T_1^k(x)) \ne \mathbf{p}^2_N(T_2^k(x)))\}) \\
&\le \frac{k}{N+k} + \varepsilon.
\end{aligned} \tag{7.44}$$

Assume we have constructed φ_r with

$$\int \bar{d}_{r(N+k)}(\mathbf{p}^1_{r(N+k)}(x), \mathbf{p}^2_{r(N+k)}(\varphi_r(x))\,\mathrm{d}\mu_1 \le \frac{k}{N+k} + 3\varepsilon. \tag{7.45}$$

To apply Lemma 7.29 again we let

$$A_j = \bigvee_{i=0}^{r(N+k)-1} (\bar{P}_j), \quad j = 1, 2$$

and

$$Q_j = \bigvee_{i=r(N+k)+k}^{(r+1)(N+k)-1} T_j^{-i}(\bar{P}_j), \quad j = 1, 2. \tag{7.46}$$

We conclude we can construct $\varphi_{(r+1)}$ without modifying the measures of sets in $A_1 \vee \varphi_r^{-1}(A_2)$ so that

$$\int \bar{d}_{N+k}(\mathbf{p}^1_{N+k}(T_1^{r(N+k)}(x)), \mathbf{p}^2_{N+k}(T_2^{r(N+k)}(\varphi_{r+1}(x)))\,d\mu_1$$

$$\leq \frac{k}{N+k} + \mu_2(\{x; \mathbf{p}^1_N(T_1^{r(N+k)+k}(x)) \neq \mathbf{p}^2_N(T_2^{r(N+k)+k}(\varphi_{r+1}(x)))\})$$

$$\leq \frac{k}{N+k} + \frac{1}{2}\sum_{\substack{a_1 \in A_1 \\ a_2 \in A_2}} \mu_1(a_1 \cap \varphi_{r+1}^{-1}(a_2))\,\|D(Q_1|a_1) - D(Q_2|a_2)\|_1$$

$$\leq \frac{k}{N+k} + \sum_{a_1 \in A_1} \mu_1(a_1)\left\| D\left(\bigvee_{i=k}^{N+k-1} T_1^{-i}(\bar{P}_1)|T_1^{-r(N+k)}(a_1)\right) - D\left(\bigvee_{i=k}^{N+k-1} T_1^{-i}(\bar{P}_1)\right)\right\|_1$$

$$+ \left\| D\left(\bigvee_{i=k}^{N+k-1} T_1^{-i}(\bar{P}_1)\right) - D\left(\bigvee_{i=k}^{N+k-1} T_2^{-i}(\bar{P}_2)\right)\right\|_1$$

$$+ \sum_{a_2 \in A_2} \mu_2(a_2)\left\| D\left(\bigvee_{i=k}^{N+k-1} T_2^{-i}(\bar{P}_2)|T_2^{-r(N+k)}(a_2)\right) - D\left(\bigvee_{i=k}^{N+k-1} T_2^{-i}(\bar{P}_2)\right)\right\|_1. \tag{7.47}$$

As $T_j^{-r(N+k)}(a_j) \in \mathscr{P}_{\bar{P}_j}$, $j = 1, 2$, expression (7.47) is at most

$$\frac{k}{N+k} + \sum_{a_1 \in A_1} \int_{a_1} \left\| D\left(\bigvee_{i=k}^{N+k-1} T_1^{-i}(\bar{P}_1)|\mathscr{P}_{\bar{P}_1}\right) - D\left(\bigvee_{i=k}^{N+k-1} T_1^{-i}(\bar{P}_1)\right)\right\|_1 d\mu_1$$

$$+ \sum_{a_2 \in A_2} \int_{a_2} \left\| D\left(\bigvee_{i=k}^{N+k-1} T_2^{-i}(\bar{P}_2)|\mathscr{P}_{\bar{P}_2}\right) - D\left(\bigvee_{i=k}^{N+k-1} T_2^{-i}(\bar{P}_2)\right)\right\|_2 d\mu_2 + \varepsilon$$

$$\leq \frac{k}{N+k} + 3\varepsilon.$$

As $\mu_1(a_1 \cap \varphi_r(a_2)) = \mu_1(a_1 \cap \varphi_{r+1}(a_2))$ we have not disturbed the relative sizes of matched names on indices 0 to $r(N + k)$. Noting

$$\bar{d}_{(r+1)(N+k)}(\mathbf{p}^1_{(r+1)(N+k)}(x), \mathbf{p}^2_{(r+1)(N+k)}(\varphi_{r+1}(x))$$

$$= \frac{r}{r+1}\bar{d}_{r(N+k)}(\mathbf{p}^1_{r(N+k)}(x), \mathbf{p}^2_{r(N+k)}(\varphi_{r+1}(x))$$

$$+ \frac{1}{r}\bar{d}_{(N+k)}(\mathbf{p}^1_{(N+k)}(T_1^{r(N+k)}(x)), \mathbf{p}^2_{(N+k)}(T_2^{r(N+k)}(\varphi_{r+1}(x))) \tag{7.48}$$

completes the induction. ∎

Proposition 7.31 *Weakly Bernoulli processes are finitely determined.*

Proof If $(\bar{X}_1, \bar{P}_1)$ is weakly Bernoulli, then for $\varepsilon > 0$, there is a k and $N > 4k/\varepsilon$ so that

$$\left\| D\left(\bigvee_{i=k}^{k+N-1} \bar{T}_1^{-i}(\bar{P}_1)|\mathscr{P}_{\bar{P}_1}\right) - D\left(\bigvee_{i=k}^{k+N-1} \bar{T}_2^{-i}(\bar{P}_2)\right)\right\|_1 < \frac{\varepsilon}{8}. \tag{7.49}$$

By Lemma 7.28, if $(\bar{X}_2, \bar{P}_2)$ satisfies

(1) $\|\bar{X}_1, \bar{P}_1; \bar{X}_2, \bar{P}_2\| < \delta$; and

(2) $h(T_2, \bar{P}_2) > h(T_1, \bar{P}_1) > \delta,$ (7.50)

then

$$\left\| D\left(\bigvee_{i=k}^{k+N-1} \bar{T}_2^{-i}(\bar{P}_2)|\mathscr{P}_{\bar{P}_2}\right) - D\left(\bigvee_{i=k}^{k+N-1} \bar{T}_2^{-i}(\bar{P}_2)\right)\right\|_1 < \frac{\varepsilon}{4}. \tag{7.51}$$

Certainly if δ is small enough

$$\left\| D\left(\bigvee_{i=k}^{k+N-1} \bar{T}_1^{-i}(\bar{P}_1)\right) - D\left(\bigvee_{i=k}^{k+N-1} \bar{T}_2^{-i}(\bar{P}_2)\right)\right\|_1 < \frac{\varepsilon}{8}. \tag{7.52}$$

Theorem 7.30 now tells us

$$\bar{d}(T_1, \bar{P}_1; T_2, \bar{P}_2) < \frac{k}{N+k} + \frac{3\varepsilon}{4} < \varepsilon. \tag{7.53}$$

Corollary 7.32 (Ornstein's isomorphism theorem, another form) *Mixing Markov chains are finitely determined, hence any two of equal entropy are isomorphic.* ■

Corollary 7.33 *Using Exercise* 7.10, *any generating partition in a mixing Markov chain is finitely determined.* ■

The chain of arguments, from the definition of weakly Bernoulli, through Lemma 7.28 to Theorem 7.30 and ending at Proposition 7.31, can be followed under weakened hypotheses. Our estimate of $\bar{d}$-distances always came by ignoring a gap in the names of length k and then matching the names without error for a length N. This was possible as the weakly Bernoulli condition allowed us to build such a pairing of names. If we were given instead as hypothesis that the conditional future distribution $D(\bigvee_{i=0}^{N} \bar{T}_1^{-i}(\bar{P}_1)|\mathscr{P}_{\bar{P}})(x)$ could be paired, for most x, with $D(\bigvee_{i=0}^{N} \bar{T}_1^{-i}(\bar{P}_1))$ with a small $\bar{d}_N$-error, this chain of argument could again be followed leading to a stronger version of Proposition 7.31. This condition, called 'very-weakly Bernoulli' turns out to be equivalent to finitely determined. The interested reader can pursue this idea further in Ornstein (1974) and Shields (1973).

There are two directions of work we should mention at this point. A tacit assumption in our proof of the isomorphism theorem was that our systems had finite generating partitions, i.e., had finite entropy. This condition can be removed and the result extended to infinite entropy systems. Here one must follow a refining tree of partitions (Ornstein 1974).

A second and extremely useful extension, by J.-P. Thouvenot (1975), involves the investigation of processes all of which contain an isomorphic copy of a fixed process $(\bar{X}, H)$, examining structures 'relative' to this common factor process. Thouvenot investigates this 'relatively Bernoulli' property of being isomorphic to a direct product of (X, H) and a Bernoulli system. He demonstrates that a theory completely parallel to the Bernoulli theory exists in this context.

Bibliography

Ahlfors, L. (1966). *Complex analyses.* McGraw Hill, New York.

Billingsley, P. (1965). *Ergodic theory and information.* Wiley, New York.

Cornfeld, I. P., Fomin, S. V., and Sinai, Ya. G. (1982). *Ergodic theory.* Springer-Verlag, New York.

Chung, K. L. (1968). *A course in probability theory* (2nd edn). Academic Press, New York.

Doob, J. E. (1953). *Stochastic processes.* Wiley, New York.

Feller, W. (1968). *An introduction to probability theory and its applications.* Vol. 1, (3rd edn). Wiley, New York.

Feller, W. (1971). *An introduction to probability theory and its applications*, Vol. 2, (2nd edn). Wiley, New York.

Friedman, N. (1970). *Introduction to ergodic theory.* Van Nostrand, Princeton.

Furstenberg, H. (1967). Disjointness in ergodic theory, minimal sets, and a problem in Diophantine approximation. *Mathematical Systems Theory* **1**, 1–49.

Furstenberg, H. (1981). *Recurrence in ergodic theory and combinatorial number theory.* Princeton University Press.

Glasner, S. and Weiss, B. (1990). Processes disjoint from weak mixing. *Transactions of the American Mathematical Society.* (In press.)

Jacobs, K. (1962). *Lecture notes on ergodic theory.* Vols 1 and 2. University of Aarhus.

del Junco, A. and Keane, M. (1985). On generic points in the Cartesian square of Chacón's transformation. *Ergodic Theory and Dynamical Systems* **5**, 59–69.

del Junco, A., Rahe, M., and Swanson, L. (1980). Chacón's automorphism has minimal self-joinings. *Journal d'Analyse Mathematique* **37**, 276–84.

del Junco, A. and Rudolph, D. (1987). On ergodic actions whose self-joinings are graphs. *Ergodic Theory and Dynamical Systems* **7**, 531–58.

Katznelson, Y. (1968). *Introduction to harmonic analysis.* Wiley, New York.

Krengel, U. (1985). *Ergodic theorems.* W. de Gruyter, New York.

Krieger, W. (1970). On entropy and generators of measure-preserving transformations, *Transactions of the American Mathematical Society*, **149**, 453–64.

Ornstein, D. S. (1974). *Ergodic theory, randomness and dynamical systems.* Yale University Press.

Ornstein, D. S. and Weiss, B. (1980). Ergodic theory of amenable group actions I, the Rohlin lemma. *Bulletin of the American Mathematical Society*, **2**, 161–4.

Ornstein, D. S. and Weiss, B. (1983). The Shannon-McMillan-Breiman theorem for a class of amenable groups. *Israeli Journal of Mathematics*, **44**, 53–60.

Parry, W. (1969). *Entropy and generators in ergodic theory.* Benjamin, New York.

Rohlin, V. A. (1966). Selected topics in the metric theory of dynamical systems, *American Mathematical Society Translations*, Series 2, **49**, 171–240.

Royden, H. L. (1968). *Real analysis* (2nd edn). MacMillan, New York.

Rudolph, D. (1979). An example of a measure-preserving map with self-joinings and applications. *J. d'Analyse Math.* **35**, 97–122.

Shields, P. (1973). *The theory of Bernoulli shifts.* University of Chicago Press.

Smorodinsky, M. (1971). *Ergodic theory; entropy,* Springer Lecture Notes 214. Springer-Verlag, New York.

Stone, D. M. (1950). Decomposition of measure algebras and spaces. *Transactions of the American Mathematical Society,* **69**, 142–60.

Thouvenot, J.-P. (1975). Quelques propriétés des systèmes dynamiques qui se décomposent en un produit de deux systèmes dont l'un est un schéma de Bernoulli. *Israeli Journal of Mathematics,* **21**, 177–207.

Walters, P. (1982). *An introduction to ergodic theory.* Springer-Verlag, New York.

Index